Art Maps and Cities

Gloria Lanci

Art Maps and Cities

Contemporary Artists Explore Urban Spaces

palgrave
macmillan

Gloria Lanci
University of the West of England
BRISTOL, UK

ISBN 978-3-031-13305-3 ISBN 978-3-031-13306-0 (eBook)
https://doi.org/10.1007/978-3-031-13306-0

This Palgrave Macmillan imprint is published by the registered company Springer Nature Switzerland AG.
The registered company address is: Gewerbestrasse 11, 6330 Cham, Switzerland

Acknowledgements

'Art maps and Cities' started as a personal journey motivated by my curiosity in exploring new subjects, but the guidance and encouragement that I received from many people along the way made a decisive contribution to its completion.

This book is based on a doctoral thesis, and I owe special thanks to my supervisors Marco Iuliano and Les Roberts, for their support throughout my time at the University of Liverpool, being always receptive to new ideas and debates. Thanks are also expressed to examiners Simon Pepper and Luci Nuti, who gave me invaluable feedback and inspiring comments that enabled me to have confidence in my work.

I am in debt with the artists who gracefully received me in their studios and offices, making time to engage in long conversations: Layla Curtis, Ben Johnson, Ruth Levene, Christian Nold, Inge Panneels, Matthew Picton, Jeffrey Sarmiento and Stephen Walter. They generously shared their knowledge, experience and original thoughts and also allowed image reproduction of their artworks. Without them this book would not be possible.

I have written large parts of this book while researching in libraries and public archives. I am grateful to the library staff at the University of Liverpool, Liverpool John Moores University and Manchester Metropolitan University for their assistance, and especially to Ludovik Sebire at the University of the West of England: in thanking him I want to extend my gratitude to all librarians. I also reserve special thanks to the staff at Liverpool Record Office, who kindly granted permission to reproduce images of their map collection.

I thank my academic colleagues who contributed with time and energy on listening and reflecting upon my writings, in particular Matthew Brown and Richard Mourne for their patient readings and inestimable help in revising drafts of this book.

Finally, I want to thank the hospitality of the people I have encountered in Liverpool, with whom I have learnt to connect with this city from the ground.

Contents

List of Figures

CHAPTER 1

Introduction

Cities are intrinsically connected with mapmaking. For as long as cities have been established, humans have been finding ways to register their morphology and location, annotating elements as physical landmarks or urban life events. Throughout history, from the pre-historic rock engravings of the first settlements to the intricate printings of the Renaissance and the current hyperreal digital visualisations, city maps have been produced as repositories of human knowledge, and mapmaking has become an authoritative tool to understand, plan and design new cities and new urban lifestyles.

The latest chapter of this long history started sometime between the 1910s and 1920s, when some artists removed maps from the usual repositories, such as scholarly books and atlases, and playfully started to experiment with them as art subjects. This was a humble beginning for what would follow, yet it was a remarkable event, even more so because in the preceding nineteenth century cartography was instituted as the major discipline of all things related to maps and mapmaking. Cities then, reaching increasing levels of densification and sub-urbanisation, were mapped under the rules of accurate surveying methods. This map production was strongly associated with another new discipline of the time: city planning. Fast forward to the 1960s, considerable debate, followed by critical practices, was directed to the dogmas of both disciplines of cartography and planning, through the emergence of movements such as 'counter-cartography' and 'New Urbanism'. Artists, again, were ahead of their

G. Lanci, *Art Maps and Cities*,
https://doi.org/10.1007/978-3-031-13306-0_1

peers and had been experimenting with interactive urban mappings as early as the mid-1950s through the situationist dérives and psychogeography.

This book reflects upon this 'last chapter' to discuss how the legacies of over a century of art maps have intruded on the realms of urban cartography and city planning. A point of depart is that the possibilities for alternative knowledges of the city are enhanced in artistic practices, not diminished. I will argue that there is a fundamental epistemological shift in including such alternative knowledges into city planning, which is much related to intuitive and humanistic approaches. Beyond measuring and quantifying the city from an abstract viewpoint, where places are coded and categorised, artists have the capacity to insinuate themselves through particularities of the urban space, relieved of the burden to commit with precise and realistic representations. That is what I am referring to as 'cartographies from the ground': maps that tell hidden stories, that register unaccounted moments, that pursue concrete, either real or imagined, viewpoints of the daily life in cities.

Another consideration in this book is to understand maps as artefacts embedded within human culture, while manifestations of ideologies, social interactions and power relations. Examining contemporary art maps and artists' practices in the urban context is to investigate the politics of space, environmental sustainability, new technologies, communities and many other approaches to urban living and urban design that are part of current debates on how to see, perceive and understand cities. And much of what is seen, perceived and understood of cities is now mediated by digital technologies of mapmaking.

The advent of mobile technologies in the late 1990s and particularly web mapping in the early 2000s gave rise to a new relationship between users and maps. At the inception of the Internet, digital maps were already available, by platforms such as MapQuest, but at this early beginning, they were static and looked like old-fashion atlases, showing one map frame at once. Things changed dramatically when Google Maps went live in 2005, providing a platform that was fully interactive in terms of scalable searches and navigation. Since then, with similar applications being developed, web mapping turned into a constant feature in the daily lives of people in all continents of the world, thanks to the expansion of satellite devices in orbit. Taken for granted and trusted beyond doubt, mapping computing technologies became pervasive, without being further interrogated by their users. Having houses, workplaces, schools, pharmacies or bus stops located instantaneously in pocket digital devices is now an ordinary, matter-of-fact feature, and little thought is given to how they 'appear' in the maps, how map applications are created and designed and, above all, by whom. This invasive and ubiquitous phenomenon has been transforming our interaction with the urban space, and in main cities all over the world, geographical information can be consumed continuously everywhere in the range of a broadband or mobile data coverage. Navigating the urban space using a smartphone and mapping software was made seamless and smooth by ever more sophisticated, user-friendly web applications.

Digital cartography associated with communication technologies significantly changed the way maps are produced and consumed and not surprisingly attracted a great interest from artists in maps and mappings. John Noble Wilford, defining what he called a 'new geography', considers that digital technologies have induced a revolution in map production and consumption that is the equivalent of that observed in the Renaissance, when new printing processes enabled the popularisation of maps and map collections.[1] Contributing to this 'map revolution' is the democratisation of geographic information systems (GIS) and open-source software, allowing non-trained and non-expert users to create customised maps, using either open institutional databases or self-gathered data. Almost at the same time that Google launched their version of online mapping, initiatives such as OpenStreetMap, a collaborative, crowd-sourcing project, were on the way to overturn the rules of 'official' cartography, creating the first free, community editable, geographic database of street networks. In their edited book of 2006 *Else/where: Mapping New Cartographies of Networks and Territories*, architecture critics Janet Abrams and Peter Hall, envisioning the potential of new technologies to unleash more creative ways of mapmaking, point out that, freed from the responsibility of accurately representing the territory, professional and amateur cartographers can provide visual representations that are more akin to other disciplines, including architecture and urban design.[2]

However, it must be considered that perceptions of urban spaces are increasingly being influenced by geolocation services that provide

[1] Wilford, John Noble (2002). *The Mapmakers: The Story of the Great Pioneers in Cartography—From Antiquity to the Space Age*. Revised ed. London: Pimlico, pp. 409–425.

[2] Abrams, Janet and Hall, Peter (eds.) (2006). *Else/where: Mapping New Cartographies of Networks and Territories*. Minneapolis: University of Minnesota Design Institute, p. 12.

narrowed selections of places that can be observed and experienced in a city. The amount and density of commercial enterprises and services, for instance, can subtly guide users to perceive more vibrant and dynamic neighbourhoods, while lack of information can denote areas 'empty' of interest. Likewise, nuanced contradictions that are part of any city are ignored, reducing the visual and informational 'noise' and making maps, supposedly, more legible. Cities are evocative spaces, as they are the bearers of memory of human settlement. 'A house that has been experienced is not an inert box. Inhabited space transcends geometrical space', Gaston Bachelard wrote in *The Poetics of Space*.[3] Applying that to the urban space is to say that a city transcends its physical form, as a space inhabited, experienced and marked with the passing generations of dwellers. What is left in the landscape that can be observed at any moment is the memory of those experiences. Conversely, the urban landscape is imagined and planned, or mapped, prior to its existence: in this case, maps are conduits to urban futures.

There is undoubtedly a general enthusiasm surrounding maps and mapping nowadays, along with concerns, criticisms and activisms. The arrival of new digital technologies has not been a hindrance to the variety and extension of artistic expressions observed in the past three decades, and artists are engaging with maps and mappings through performance, psychogeography and locative media, producing site-specific and virtual artworks and being critical of traditional cartography and its methods of representing space. The contemporary ways that art maps and artistic practices can be used to understand and relate to cities is a central object of discussion of this book, and within this premise, artworks are approached as creative responses to the perception of the urban environment. Rather than debating intrinsic aesthetic qualities of artworks, I am considering them as translations, or modes of interpretation, of cities. In looking at art maps and artistic mapmaking, considering both artefacts and associated process, I seek to investigate alternative cartographies of urban spaces that go beyond the production of maps and plans intended to perform specific tasks of information, navigation, planning or design. This also implies to discuss what kind of knowledge artworks can produce and communicate, what meanings art maps convey and how artistic mapmaking can constitute a method for interpreting urban spaces.

[3] Bachelard, Gaston (1969). *The Poetics of Space*. Trans. Maria Jolas. Boston: Beacon Press, p. 47.

Definitions of maps and mappings often fall under the 'art' and 'science' binary, unfolding into systems of knowledge that consider 'alternative' and 'conventional' cartographies, depending on 'qualitative' and 'quantitative' methodological approaches. Those have parallels with the definitions of what a city is and what constitutes an urban environment: there are interpretations linked to the idea of a city as a complex mechanism to be efficiently operated and others more akin to humanistic views of social and cultural complexity. Both cartography and urban design are commonly understood as subjects that combine art *and* science, without specifying further what concepts make these categories as separated realms of knowledge. In a simplistic definition using pair of opposites, art is knowledge that is 'invented', while scientific knowledge is 'discovered', a polarity constructed along the twentieth century as an 'economy of the binary'.[4] Under this perspective 'art map' is either a contradictory category or an inventive appropriation of the science of cartography. Such approaches are not very helpful when analysing artistic mapping practices, especially in the contemporary world.

An interesting approach to the question of map 'categories' was suggested by philosopher Edward Casey, who identified 'four ways to map'.[5] In the first, *mapping of*, the aim is to capture the exact geographical features of a place, through the means of surveying, positioning and scaling the territory in the context of the globe. The second, *mapping for*, relates to maps that are made for particular purposes, for example, to illustrate tourist guides, and do not have a commitment to cartographic accuracy. The third category, *mapping within/in*, concerns the way one experiences a place, and the purpose is not to present it as an accurate account or a piece of information but rather to communicate a feeling, and the viewer 'is drawn *in* rather than encouraged to hover *above*'.[6] The final category, *mapping out*, relates to a type of mapping that extracts its matter from the land, one that is not only about 'tracing out the mere outlines of objects in the landscape' but that also follows 'the lineaments of the earth, putting down its main marks, limning its first features'.[7]

[4] Jones, Caroline A.; Galison, Peter (eds.) (1998). *Picturing Science, Producing Art*. New York: Routlegde, p. 2.

[5] Casey, Edward (2005). *Earth-Mapping: Artists Reshaping Landscape*. Minneapolis: University of Minnesota Press, pp. xx–xxi.

[6] Casey, Edward. (2005). Op. cit., p. xxi. (in italic in the original).

[7] Casey, Edward. (2005). Op. cit., p. xxi.

On proposing this 'taxonomy' of mapping practices, Casey was interested in building an argument for land art outlining progressive levels of embodiment with the landscape, from the detached 'mapping of' experience to the more intimate 'mapping out'. It is not difficult to associate such approach with the binary science/art, and the categories can be viewed as a spectrum from 'scientific' to 'artistic' ways of mapping. However, in analysing contemporary art mappings I would argue that the four categories can be found in artists' practices, and what sets apart 'artistic' mapping from 'scientific' mapping is that artists do not aim to map places with geographical accuracy *only* (or not at all) but to experiment with accuracy, if it is the case, and moreover to express and communicate ideas other than cartographic precision. Mapping for artists is a tool, a metaphor, a trope to address and comment on place and space, visual representation and art itself. Through experimenting and finding new interpretations in map form, they are able to engage with the production of scientific knowledge in critical and creative ways.

The roots of what I call in this book contemporary art mapping can be traced back to the early twentieth century, as mentioned earlier, and the first experiments in this field were conducted by surrealists and dadaists. However, it was only in the mid-1950s, with the emergence of the Situationist International, that mapping acquired a new meaning, especially in the context of the urban environment, associated with psychogeography. That continued throughout the 1960s and 1970s, in art movements like conceptual art and land art. Concurrent with this period was the development of digital cartography, through GIS. These two apparently separated events—the changes in the arts and in informational science—would converge in new practices from the mid-1990s onwards, such as 'radical cartography' and 'counter-mapping', that merged the 'four ways' of mapping described by Casey.

All maps face the challenge of interpreting what is observed in the real world, but unlike scientific traditions of mapping, expected to thoroughly represent the world 'as it is', art mapping has the potential to unveil what is seen across. And for that purpose, art maps are allowed to be partial, selective and not all-encompassed, interpreting (not necessarily 'representing') the world 'as it could be', or 'could have been'. Artists can let themselves be equally guided by intuition as well as deduction, to be open to the production of different kinds of knowledge other than based on accuracy and precision, which are qualities sought after in cartographic science. Art maps offer a double translation act: first from reproducing a reality and second on reinterpreting this reality, giving new meanings, obliterating, emphasising, retelling stories, creating new categories. While art maps do not bear the commitment of being functional tools, such as for navigation and wayfinding, they can operate as creative devices for experimentation.

It is intended in this book to present the diversity of theories, ontologies and epistemologies that surround maps and mapping practices, which are useful to discuss the realm of arts in context. Understanding the theoretical context will also allow us to discuss how artistic and scientific approaches to maps and mapping are not exactly opposites but dialectical partners, a question explored here in the study about art movements and in the artists interviewed for this project.

Considering urban art maps as translations of cities, I intend to investigate map theories elaborated from the 1990s onwards, such as the claim for the agency of mapping,[8] the performative character of maps and mapping[9] and the possibilities of deep mapping,[10] relating them to urban theories and more specifically to the conceptual triad developed in Henri Lefebvre's work *The Production of Space*, first published in 1974.[11] Lefebvre sought a conception of the urban space beyond the binary of the space 'perceived' and the space 'known' with his proposition of the 'lived' space: that of human practice and experimentation. Likewise, I argue here that art maps offer the opportunity to analyse space beyond the dialectics of 'seeing' and 'registering' (or representing) the world to 'translating' it, communicating not one reality (that of a map that is an accurate version, or mimetic version of that world seen) but multiple realities (that of the performative character to the map).

[8] Corner, James (1999). The agency of mapping: speculation, critique and invention. In Cosgrove, Denis (ed.) *Mappings*. London: Reaktion, pp. 213–252.

[9] Cosgrove, Denis and Martins, Luciana (2000). Millennial geographics. *Annals of the American Association of Geographers*. 90 (1) pp. 97–113; Crampton, Jeremy (2009). Cartography; performative, participatory, political. *Progress in Human Geography*, 33 (6), pp. 840–848.

[10] Bodenhamer, David; Corrigan, John and Harris, Trevor (eds.) (2015). *Deep maps and spatial narratives*. Bloomington: Indiana University Press; Wood, Denis (2015) Mapping deeply. *Humanities*, 4(3), pp. 304–318; Roberts, Les (2015) Deep mapping and spatial anthropology. *Humanities*, 4(3) (editorial).

[11] This corresponds to the French edition; the first English translation only appeared in 1991. Lefebvre, Henri (1991). *The production of space*. Trans. Donald Nicholson-Smith. Oxford: Blackwell.

Cities were a central subject in Lefebvre's inquiries, and the use of his theoretical framework[12] is useful for establishing connections between contemporary art practices of mapping and urban spaces. Lefebvre also engaged with the politics of urban space, being a collaborator as well as a mentor for the Situationist International (SI), a group of artists and intellectuals formed in Paris in the 1950s that explored maps and techniques of mapping as tools to challenge traditional perceptions of urban spaces. His scholarship is frequently related to the emergence of critical urban theories in the late 1960s, which had contributions from leftist or radical urban thinkers.[13]

The mapping artworks analysed in this book as a sample of the contemporary art map practices were all produced around the mid-1990s and after. They emphasise experimentation alongside experience, meaning defiance of cartographic codes and commitment to different ways of mapping. This does not mean that contemporary artists are taking a totally new approach, but it represents a step further into exploring mapping practices as means to visualise, interpret and critique urban spaces. Nonetheless, one must question how much of these 'urban spaces' depicted in the art maps are representative of cities around the world and of urban cultures that shape spaces in a variety of landscapes.

Researching for this book it soon became clear that most of references, in the theoretical framework and artistic examples, were from British and American sources, with the few exceptions of French philosophers and Japanese artists. The literature reviewed is overwhelmingly anglophone, and I am also aware of the striking preponderance of Anglo-American male authors. This is related to ways that knowledge propagated in the fields of geography and city planning, as well as in visual arts practices, in the twentieth century. Firstly, language counts—and English became the lingua franca after WWII in politics, media outlets, technology, academic circles and the art world. Secondly, intellectual and artistic networks are manifested in physical spaces, and geographical proximity matters. Cities are privileged spaces for nurturing such networks, concentrating research and development, attracting academic workforce and providing spaces for

[12] Sometimes referred as 'spatial trialetics'. The term was not used by Lefebvre but proposed by American geographer Edward Soja, who further developed Lefebvres's concepts. Soja, Edward W. (1996). *Thirdspace: journeys to Los Angeles and other real-and-imagined places*. Oxford: Blackwell.

[13] Brenner, Neil (2009). What is critical urban theory? *City*, 13:2–3, pp. 198–207, p. 198.

socialisation. Capital cities and other major urban centres located in Europe and the USA offered the best opportunities for such 'intellectual friction' and social interaction in the first half of the twentieth century.

Since the 1960s this scenario started to change, at the time of the emergence of conceptual art and critical theories that addressed the impact of dominant cultures and identities, and such change is accelerated in the twenty-first century. The de-colonisation of history and geographical knowledge is an ongoing process that is bringing significant consequences to scholarship. In relation to maps, as more non-Western, non-Eurocentric histories are researched and disseminated, this process opens out to a fascinating opportunity for scholars and artists alike to explore a much wider world, in its spatial and temporal dimensions.

Despite a long history of representations of cities in maps, and the relationships that can be envisaged between art maps and cities in the contemporary world, there is limited research on how artists intervene in the realm of urban cartography. Focusing on graphic forms of mappings produced by visual artists seemed to be a good starting point of investigation. The reason for that is twofold: on the one hand, the immense variety of mapping methods that I encountered during my research was overwhelming, and it would not be useful to exhaustively examine them, as it was not intended to produce a compendium of art practices. On the other hand, graphic maps are the most likely used in city planning, and the correspondence of one practice with another allowed me to discuss how traditional cartographies can benefit from artistic mappings but also to point out that much of experiment and innovation in mapping found in avant-garde urban and architectural design throughout the twentieth century and beyond are predominantly based on visual media.

The research undertaken for this book was developed from both my long-established research interest in the urban landscape and more recent training and work experience in geographic information systems. Although these subjects were pursued in quite different contexts of knowledge—the former in the humanities, following an academic background in architecture and planning; the later within the science and technology of spatial analysis and database management—they came together unexpectedly in a project combining visual arts, mapping and cities. The methodological approach taken in my research derives from this interdisciplinarity, and the book is exploratory and interpretative in nature, which means that it does not intend to test hypotheses, or to strictly follow singular systems of thought, but rather to investigate the confluence of theories originating in different domains in a process of discovery of what creative artistic practices can bring into the understanding of urban spaces through mapping. Therefore, many disciplines are involved in this process of personal discovery, as the authors cited in this book will indicate, from urban geography, philosophy, visual arts, cartography and urban sociology.

The research is supported by qualitative methods, through in-depth interviews with eight visual artists based in or working in the UK, investigating their techniques, concepts and personal experiences in mapping and mapmaking. Among these artists are the four authors of three maps produced for Liverpool, England, between 2005 and 2011, which constitutes our case study examining different artistic mapping practices in the same urban and historical context. It is important to highlight here that this book does not intend to engage in a fundamental critique of art maps but to examine how and why they are produced, thus justifying the adoption of interviewing as method.

The interviews were semi-structured; there were no pre-arranged questions, no time limits or other restrictions that could have prompted controlled responses. I asked the artists about their personal motivations for using cartography in their practices, their methods and techniques, their artistic references and their own worldview in relation to space, politics, art and society. The interviews turned into informal conversations and sometimes run well over the allocated time. There was in effect an exchange of information and moreover 'an interchange of views', sharing knowledge over themes of mutual interest.[14] Talking with artists in this way also proved to be a useful method to grasp the amplitude of their discourses regarding their own practices, instead of approaching them with pre-conceived labels or categories on the attempt to reach a 'common' ground in our dialogue. As Tim Griffin, former editor-in-chief of the international contemporary art magazine *Artforum*, commented on his own interviewing experiences with artists:

[14] Brinkman, Svend (2014). Unstructured and semi-structured interviewing. In Leavy, Patricia. *The Oxford handbook of qualitative research*. New York: Oxford University Press, p. 278.

> An interview works best when it inspires in artist and interviewer a kind of mutual disequilibrium, an active and continuous revaluation of the circumstances in real time.[15]

The artists interviewed for the project are near contemporaries, from a generation in its majority born after 1970 who engaged with mapping in their professional careers in the noughties. Except for one, all artists were based in Britain at the time of the interviews, which made it possible to have face-to-face encounters and to facilitate visiting artists' studios, enriching the experience of the interview, giving an insight on the physical space where they create their artworks and allowing a deeper interaction between us, as they could talk about past and present projects while showing examples of their work, tools and techniques. Four artists with quite distinct practices were interviewed to offer a general view of the status of contemporary art mapping in Britain and from British artists: Matthew Picton, Ruth Levene, Layla Curtis and Christian Nold. Picton produces paper sculptures based on cartographic urban plans, usually focusing on a historic narrative. He was the only artist interviewed over the Internet, as he lives and works in Oregon, USA. Ruth Levene is based in Sheffield and her work involves performance, video installations, digital drawings and participatory events. Layla Curtis is based in London and produces collaged and hand-drawn maps, besides working with video combined with digital platforms. Christian Nold, also based in London, is an artist and researcher who works in participatory models of mapping cities. The other four artists interviewed are authors of three maps for Liverpool: Ben Johnson, Stephen Walter, Inge Panneels and Jefrey Sarmiento. Johnson is a painter who produces large-scale cityscapes and architectural spaces; he lives and works in London. Stephen Walter, also from London, is a printmaker and works on hand-drawn maps and landscapes. Inge Panneels and Jeffrey Sarmiento are glass artists and were based at the National Glass Centre, in Sunderland, at the time of the interview.

Even if not originally from Britain (Inge Panneels is from Belgium and Jefrey Sarmiento from the Philippines) or not based in Britain (Matthew Picton lives and works in the USA), all the artists interviewed were faced

[15] Griffin, Tim and Anastas, Rhea (2005). Method acting: the artist-interviewer conversation (with response). *Art Journal*, vol. 64, n. 3, pp. 70–83.

with a strong cartographic tradition that is much engrained in the British imagination and culture in producing their art maps. The UK, alongside France,[16] was one of the first countries in the world to employ a systematic territorial survey, using the technique of triangulation, leading to the creation of the Ordnance Survey (OS), Britain's mapping agency, in 1791.[17] The OS was an enterprise founded in the spirit of the Enlightenment, in pursuit of creating 'an accurate image of the natural world',[18] and became an icon of national identity. This official, science-based cartography would also support the construction of a sense of power and domination exercised by the leading countries in the colonial expansion of the eighteenth and nineteenth centuries—France and Britain. Most of the artists interviewed have used in their practices OS cartographic products, on paper and digital formats, for geographical reference and as a base or graphic 'background canvas' onto which they overlay their own artwork. In some cases, they have also produced artworks by defacing or modifying information from OS maps, or manipulating geographical data, in an act of questioning the imperatives of cartographic rules and the institutionalised practices of mapping. Some artworks involve public participation; others are inspired by the public, as well as from local cultures and stories. They are often produced in a studio, supported by archival research and secondary data, but mapping artists rely much on field work, visiting sites, talking with people and experiencing the territory by themselves. Their art influences span from modernist and contemporary painters to the situationists, land art and conceptual artists. Digital technologies are tools employed indirectly and the level of abstraction varies; some pieces appear more 'mappy' (borrowing a word from one interviewed artist) than others, meaning they resemble more traditional maps.

[16] The first thoroughly surveying, the 'Principal Triangulation of Great Britain', was undertaken between 1791–1853.The 'Cassini Maps', the first national maps of France, were produced between 1750–1815.

[17] The creation of the OS has military roots, as its name indicates ("ordnance" is a synonym of "mounted guns" or "artillery"), emerging from a military operation to map the country after the Jacobite rising in 1745. The first printed map appeared in 1801 and the first series of maps in 1870. In 1995 OS launched its website and finished the digitalisation of its paper maps, making Britain the first country in the world to complete a programme of large-scale electronic mapping. (https://www.ordnancesurvey.co.uk/about/overview/history.html). Nevertheless, the most loved paper maps series are still being produced and in use, despite the pervasiveness of the digital map.

[18] Hewitt, Rachel (2010). *Map of a nation: a biography of the Ordnance Survey.* London: Granta Books, p. 113.

The case of Liverpool is relevant in this context because being a wealthy port city in the past, it was extensively mapped and documented. After a period of sharp economic decline during the 1980s, with some dramatic social consequences, such as job losses, riots, poverty and deprivation, a recovery process started to become more visible in the early 2000s with improvements in its business centre, cultural venues and public spaces. The city's physical and social renaissance was encapsulated by a celebration of its history, culture and people in 2008 as European Capital of Culture, and to mark the occasion, two art maps were commissioned: a cityscape painting (by Ben Johnson) and a glass sculpture (by Panneels and Sarmiento). These, alongside a hand-drawn map (by Stephen Walter) produced around the same time, constitute the three pieces that are examined as a case study.

The book is structured according to a concept of moving from the general to the particular, which is reflected in the order of chapters and the presentation of the interviews and case study. Thus, Chap. 2 elaborates on a theoretical discussion based on three core concepts in this project: those of 'translation', 'urban' and 'contemporary art', seeking to identify a conceptual framework of 'urban art map' exploring related fields of knowledge within the cultural contexts of the language, the urban narratives and the art world. Chapter 3 ventures into epistemological perspectives of mapmaking, approaching the nature of seeing, representing and performing: firstly, by examining the dialectics of the views 'from above' and 'from the ground' in the urban context, through the works of phenomenological scholars and in the philosophical approach of the Western 'visual paradigm' and, secondly by discussing modes of representation in cartographic traditions in relation to performance, agency and deep mapping. Chapter 4 presents a historical (and chronological) perspective of the contemporary art map, from its foundations in the Surrealism and Dada art movements in the 1920s, to the situationists and conceptual artists of the mid-twentieth century, to the counter-mappings of the early twenty-first century. Chapters 5 and 6 examine in more detail examples of contemporary art maps and mappings with emphasis on the urban theme, exploring the contents of the interviews and the case study. Chapter 5 introduces the more recent production of contemporary artists in the past two decades, first presenting four interviews with British artists according to emergent themes and then analysing five artworks that explore the urban pattern.

Chapter 6 introduces Liverpool as a case study through its cartographic history up to the twenty-first century, followed by a description of the events that led to the production of three art maps for the city between 2005 and 2011, and ending with the interviews of the artists involved in this process. Chapter 7 concludes the book, summarising the main discussions in the face of the theoretical framework adopted and discussing the significance and limitations of the research project.

CHAPTER 2

Urban Art Maps in Cultural Context

The emergence of critical discourses in geography in the 1980s concerning the methods and practices of cartographic representations of spaces generated a wealth of literature on the nature of maps and the ontological and epistemological underpinnings of maps and mappings. Recently a greater attention has been given to mapping practices and their performative character in scholarly research.[1] However, there is a lack of a more focused investigation on the confluence of artistic mapping practices and the urban realm. Under this perspective, and to provide a cultural context underpinning practices in contemporary art that focus on mapping cities, I propose a straightforward theoretical framework through the analysis of three basic concepts, highlighting their intersections: that of 'translation', 'urban' and 'contemporary art'. Each of them instigates a discussion around interlinked definitions, allowing to explore the 'urban art map' as a category across diverse disciplinary fields.

The first concept, 'translation', explores the notion that maps, as art, are expressions to communicate and convey meaning, and defining what a map is implies to understand the mapping practice that produces it as acts of 'translating' spatial knowledge. Therefore, the definition of map is historical, circumstantial and subjective. From this premise I discuss ideas of

[1] According to Kitchin et al., this is a trend that has been in evidence since 2007. Kitchin, R., Gleeson, J. and Dodge, M. (2013). Unfolding mapping practices: a new epistemology for cartography. *Transactions*, 38, 3, pp. 480–496.

G. Lanci, *Art Maps and Cities*,
https://doi.org/10.1007/978-3-031-13306-0_2

visual representations in the cultural contexts of the language, linking to the second concept, 'urban', which is presented as narratives of the city. I discuss social theories that oriented studies in urban planning and design in the past century and more recent discourses related to smart cities, to examine definitions of city in the contemporary world. This leads to the concept of 'contemporary art', discussed as a social construct that is intrinsically connected with the 'art world' that emerged in the second half of the twentieth century, a time of increasing urbanisation that is marked by the emergence of global cities and networked societies.

Understanding Maps as Translations

Maps in their visual form combine words and images to inform about and to give sense to a specific location. They are primarily a tool for communication that prompts narratives and accounts of a place. When 'reading' a map one is seeking for meanings that unveil such narratives: we reason about distances and land shapes but also about history and culture to find out what that place is, or was, or could be, why such place is plotted in a map and how it is related to other places. Considering the textual nature of maps implies to understand them as constructs derived from codes and systems that are built to convey ideas. If maps are 'legible' it is because they use a coded language that allows significations to be extracted from them. However, language can conduct specific meanings that are 'untranslatable'; this is easier to verify when one attempts to translate a word that seems to 'only exist' in a language spoken in a very determined place or location. Slang is a good example; it changes temporally and geographically, with meanings tightly connected to certain social groups sharing the same experiences in the same place. So how to situate a 'language of maps' and its cultural meanings?

Twentieth-century linguistics has its roots in the work of the Swiss linguist Ferdinand de Saussure (1857–1913), who established the principles of semiology as a systematic study of signs, and the American philosopher and scientist Charles Sanders Peirce (1839–1914), who proposed a triadic semiotic model to explain the relations between signs, objects and the mind. Such theories, framed in the structuralism paradigm,[2] were later developed in the 1960s onwards by poststructuralist and postmodernist scholars and widely applied in arts and humanities subjects, from architecture to music and advertising, and cartography was not left behind. The French cartographer and geographer Jacques Bertin was one of the first theorists to synthesise principles of graphic communication used in cartography with the system principles of semiology. His book *Semiology of Graphics*, published in 1967, became a reference to the study of diagrams, networks and maps, providing the foundations for what is now recognised as information or data visualisation.[3]

The French theorist Roland Barthes, who wrote extensively on the subject of semiotics, observed that 'scientific geography, and in particular modern cartography, can be considered a kind of obliteration, of censorship that objectivity had imposed on signification'.[4] American geographers and map theorists Denis Wood and John Fels, influenced by his work and of other semiologists, explored the semiotic analysis of maps in an extensive article published in 1986, identifying five 'codes of extra signification': iconic, linguistic, tectonic, temporal and presentational.[5] The 'iconic code' refers to the themes or 'events' that the map depicts, associated with a certain location; the 'linguistic code' refers to toponomy or labels used in the map, which classifies and identifies the 'events'; the 'tectonic code' relates the 'events' with topological features at a certain scale; the 'temporal

[2] Saussure is regarded as the founder of structuralism, which has a systematic approach to phenomena. Language is thus considered a system of signs defined by their internal structure and their relationship with other structures. *Semiology* and *semiotics* have been used as interchangeable terms, although *semiology* is often related to the linguistic tradition initiated by Saussure, while *semiotics* is referred to a more general theory of signs based on Pierce's work. Noth, Winfried (1990) *Handbook of semiotics.* Bloomington: Indiana University Press, pp. 12–14.

[3] Bertin, Jacques (1983). *Semiology of graphics: diagrams, networks, maps.* Madison: University of Wisconsin Press. The American statistician Edward Tufte is one of the pioneers of information design and visualisation, whose book *The Visual Display of Quantitative Information* (Cheshire, CT: Graphics Press, 1983) was a breakthrough in the field; following him the Spanish information designer Alberto Cairo took data visualisation into journalism and art; one of his main publications is *The Truthful Art: Data, Charts, and Maps for Communication* (Berkeley: New Riders, 2016).

[4] Barthes, Roland. (1967) Semiology and the urban. Lecture given on 16 May 1967. In Leach, Neil (1997). *Rethinking architecture: a reader in cultural theory.* London & New York: Routledge. pp. 158–164, p. 159.

[5] Wood, Denis and Fels, John (1986). Designs on signs / myth and meaning in maps. *Cartographica*, 23 (3), pp. 54–103.

code' indicates the time and duration of events; and finally the 'presentational code' covers design features such as title, legend, insets, notes and so on.

This might appear a systematic and exhaustive way to refer to maps, based on a theoretical framework that is closer to the positivist tradition than to poststructuralist approaches that criticise semiotics as a rigid system of interpretation of reality, disregarding the dynamics of time and place, but it launched a quest for the demystification of maps and unveiled the process of creating meaning through design, using a theoretical framework that was originally conceived to analyse discourse and language.[6] One great thing that maps provide is to enhance the symbiosis of words and images, or verbal and visual codes, and a lot of what artists do is related to breaking, subverting and defying such codes, re-translating meanings first imposed by traditional cartographers.

When mentioning modern cartography Barthes was making a case for a semiotics of the city that privileges the process of uncovering of meanings of the urban space, for 'the city is a discourse, and this discourse is truly a language'.[7] Barthes advocated that 'it is not so important to multiply the surveys or the functional studies of the city, but to multiply the readings of the city', supporting an alternative approach to urban space, moving away from its scrutinisation to its interpretation. He implies that urban planners would benefit from learning to 'read the city' and that what can be mapped objectively might overlook the subjective significations that can speak for what a city is. Understanding urban maps as translations of urban spaces can contribute to unleashing the 'language' of the city, reconstructing its code, and then, as Barthes suggests, turn to 'means of a more scientific nature' keeping in mind that whatever defines urban structures ('signifiers') are historically 'vague, dubious and unmanageable'.[8]

Definitions of 'map' and 'cartography' developed by critical theorists, from the 1980s onwards, indicate how linguistic components are brought at the core of the production and use of maps and that their meaning is derived from them, through a process of interpretation:

[6] Reviewing their own article 25 years later Wood and Fels recognise a mutual enthusiasm for semiology and how Roland Barthes, Charles Pierce and Ferdinand Saussure, among others, were influential to their work. Wood, Denis and Fels, John (2011). Reflection Essay: Designs on signs/myth and meaning in maps. In Dodge, Martin (ed.). *Classics in cartography: reflections on influential articles from Cartographica*. Hoboken, NJ: Wiley-Blackwell.

[7] Barthes, Roland (1967). Op. cit., p. 160.

[8] Barthes, Roland (1967). Op. cit., p. 164.

> Far from holding up a simple mirror of nature that is true or false, maps re-describe the world – like any other document [...] within the frame of one map there may be several texts – and intertextuality' that has to be uncovered in the interpretative process.[9]

When Harley regards maps as documents that 're-describe the world', he is indicating that as tools of knowledge they imply translation: the map results from transferring what we see and perceive of the world into another medium, converting meanings/perceptions from one language to another. In doing so, we are in fact convening/rendering our understanding of the world, giving new meanings, because a translation is never the original book; it is a new book.

The historian Christian Jacob also underlines the aspects of language involved in the production of maps:

> The cartographer creates a world: not the natural world, but a cultural world, invested by one language among other possible ones, attesting to an organized space, punctuated with meaningful and constructed places, invaded by a reticulation of proper names that bear witness to the appropriation of space through chains of metaphors, fields of knowledge, components of individual or collective mythology, and the declension of lexical variations.[10]

When mapping urban spaces artists are dealing with many *versions* of the cities they depict through the type of research they do, the references they select and the medium they use to observe and perceive their subjects. This translation is in its conception and *interpretation* rather than a mimetic representation and a *transformation* of whatever materials and data the artists collect. Translations bear a transgressive characteristic: they defy authorship and authority. In the publishing industry, translators are not considered co-authors, but they can get access to royalties, besides the payment for their work, as they often work closely with authors in re-creating the original text in another language. Translating is thus an activity that infringes the borders of the 'true/original' and the 'false/copy',

[9] Harley, John B. (2001). *The new nature of maps: essays in the history of cartography*. Edited by Paul Laxton. Baltimore and London: John Hopkins University Press, pp. 35–38.

[10] Jacob, Christian (2006). *The sovereign map: theoretical approaches in cartography throughout history*. Transl. Tom Conley (first published in French in 1992). Chicago and London: University of Chicago Press, p. 206.

and in breaching those, it produces new forms of interpretation. Art maps, as we will see in this book, typically challenge traditional cartographies exerting creative transgressions akin to the practices of translators.

A translation is always a quasi-similar, articulating difference. It does not reproduce the original; it is not a copy but a re-invention. Walter Benjamin, in his essay *The task of the translator*, regards the translator as someone who needs to find an 'intended effect – intention – upon the language into which he is translating which produces in it the echo of the original'.[11] Benjamin's remark is specifically addressing literature translations, but we might consider here that the reference to the *intention* is a call for an imaginative act that nevertheless might 'echo' the original. The translator is therefore a communicator above all, bridging author and reader, functioning as a medium through which an art form is propagated. Translations can be multiple (as echoing the original), and in this sense, they allow continuous interpretations of the world.

Translating, according to the scholar Sarat Maharaj, presents an ontological paradox: on aiming to perfection, it risks turning its production into the original source and away from a translation.[12] Maharaj considers that in contemporary art practices there are attempts to capture the turbulence of reality—the 'semi-darkness of the untranslatable'—producing a 'nebulous and unstable knowledge' that cannot be apprehended by scientific enquiry. An interesting observation comes from the American translation theorist Lawrence Venuti, who considers that translating induces an 'act of violence':

> [...] the violence that resides in the very purpose and activity of translation: the reconstitution of the foreign text in accordance with values, beliefs and representations that pre-exist it in the target language, always configured in hierarchies of dominance and marginality, always determining the production, circulation, and reception of texts. Translation is the forcible replacement of the linguistic and cultural difference of the foreign text with a text that will be intelligible to the target language reader.[13]

[11] Benjamin, Walter (1999). *Illuminations: essays and reflections.* Ed. Hannah Arendt, trans. Harry Zohn. London: Pimlico, p. 77.

[12] Velasco, Alejandro del Pino (2008). Summary of 'An unknown object in uncountable dimensions: visual arts as knowledge production in the retinal arena', a presentation by Sarat Maharaj. Hlavajova, M. Winder, J and Choi, B. (eds.). *On knowledge production: a critical reader in contemporary art.* Utrecht, BAK (basis voor actuele kunst), p. 138.

[13] Venuti, Lawrence (1995). *The translator's invisibility: a history of translation.* London: Taylor & Francis, p. 18.

In cartography, knowledge is often deployed based on what can be 'seen': after all, cartography aims to turn geographical knowledge visible, to make the spatial patterns and correlations that characterise a place or territory apparent. Vision, however, is an active process and requires engagement between what is being observed and the observer. Hence, 'reading a map' is a common term referring to one's ability to understand the physical world through 'map language'. Graphic elements, legends, text and colours come together to compose a language that aims to be universal, attempting to be legible to the possible largest number of people—or so this is the aim of powerful software companies in providing standards of legibility and data compatibility. Such standards might bring a common ground for generating and communicating digital geographical data, but the overall attempt to create a universal cartographic language is perhaps illusory, and that is because even in dealing with science, translation is necessary.

On producing meanings in another language, a translator provides a version of the original or 'foreign' text, as Venuti calls it. In cartography, the mapmaker provides a version of the world observed, according to cultural conventions and historical contexts to where both map maker and map reader belong. Universality can be just considered a myth when translation imposes on the way of communicating spatial perceptions and understandings.

Art mapping as an act of translation implies performing an act of betrayal, as to translate one needs to break away from the original (to be able to make a 'readable', coherent version of it). As made commonplace in Italian language, the translator is a traitor: *traduttore, traditore*. In between the original and the translation, there will be necessarily noise, the fuzzy area where information is in its fluid state. The philosopher Michel Serres has dedicated his work to understanding communication between knowledge domains, and between artistic and scientific disciplines. Serres recognises that this communication is 'noisy', but noise is inherent to trigger the transformation necessary to advance knowledge:

> Such communication [dialogue] is a sort of game played by two interlocutors considered as united against the phenomena of interference and confusion, or against individuals with some stake in interrupting communication. These interlocutors are in no way opposed, as in the traditional conception

of the dialectic game; on the contrary, they are on the same side, tied together by a mutual interest: they battle together against noise…[14]

Maps tell stories and art maps retell them, enhancing personal or collective experiences. It is the artist's choices that matter rather than a functional impersonal narrative, choices that are primarily guided by creativity and imagination. Artists as map makers provide translations of the territory. In the process of making, they are transforming information of all kinds—geographical, historical, physical and visual—as to establish communication, transmitting that information imbued with perceptions, concepts, ideas and aesthetic choices. I am suggesting here that the artistic map is an interpretation of that noise in the sense that it reflects what the artist receives and selects as information that is going to be passed on using cartographic language. What is to be mapped? What stories does the map tell? How to convey the message to audiences? The question of communicating through maps thus becomes essential to artists, who are not alone in the struggle: when making a piece, they are attentive to the input from other 'interlocutors', being the general public they talk with to collect impressions of the place to be mapped, and also in conversation with other artists, technicians involved in the making, curators responsible for exhibitions, sponsors and commissioners, librarians and archive holders. The whole process of creating, producing and exhibiting or disseminating an artwork depends upon building an intricate communication network, where translation plays a central role.

In another take on linguistic metaphors surrounding maps and mappings, Denis Wood differentiates 'mapping' and 'mapmaking', comparing the first to 'speaking' and the later to 'writing'.[15] Wood considers that mapping is a faculty inherent to all human beings while mapmaking is a skill achieved throughout human history of *inscribing* maps: while mapping is related to narrative, mapmaking is related to registering the narrative. However, as Wood suggests, such a difference is a very fine one, just like speaking and writing. Both are acts of communication, but their social and political implications can be very different, as 'it is the inscriptive property of the artefactual map that permits it to serve the interests of

[14] Serres, Michel (1982). *Hermes: Literature, Science, Philosophy*. Edited by J.V. Harari and D.F. Bell. Baltimore: Johns Hopkins University Press, pp. 66–67.

[15] See Wood, Denis (1992). *The power of maps*. New York: Guilford Press and Wood. Denis (1993). The fine line between mapping and mapmaking. *Cartographica*, 30: 50–60.

power elites who control the mapmaking process, as well those who would contest them'.[16]

The writer and literary theorist Maurice Blanchot, in an interesting approach to the act of translating, reminds us that language is neither immutable nor static either in original or translated texts:

> The original is never stationary, and all the futurity that is contained within a language at a given moment [...] asserts itself in the solemn drift and derivation ['dérive'] of literary works. Translation is bound up with this process of becoming [...] indeed, translation is only made possible by this movement and life that it lays hold of, sometimes to release it purely, sometimes to capture it at great pains.[17]

Blanchot's observations lead to considerations on whether a city is a reality to be translated as if there is one 'original text', or why a city is intrinsically a space of transformation. If in cartography urban maps are to be 'inferior' or 'untrustworthy', because they can admittedly never represent the whole with accuracy (the projection needed to compose a map is a distortion of reality per se), in art those maps can become translations, not inferior or superior, but different views that incorporate that essential nature of urban transformation, and in doing so they become attuned with the 'original', by the wonders of playing on difference:

> Translation is the sheer play of difference: it constantly makes allusion to difference, dissimulates difference, but by occasionally revealing and often accentuating it, translation becomes the very life of this difference, and it is under the spell of this difference that translation discovers its august duty whenever it proudly sets out to bring the two languages close together through its own power of unification.[18]

Considering these thoughts map artists can be seen dealing with the cartographical 'speaking' and 'writing' on mapping and mapmaking, translating narratives and inscribing them in creative ways that divert, transgress and modify information. Because art maps are not committed to functional aspects of cartography, such as wayfinding and accurate location, they are comparable to poetic language, in the sense that art maps,

[16] Wood, Denis (1993). Op. cit., p. 53.
[17] Blanchot, Maurice (1990). Translating. Trans. R. Sieburth. *Sulfur*, (26) 82–86, p. 84.
[18] Blanchot, Maurice (1990). Op. cit., p. 83.

like poetry, can induce their 'writers' and 'readers' to feel language as contingent and necessary at the same time.[19] Textual and graphic elements in an art map come together to express meanings that are both necessarily descriptive and accidentally incomplete, open to interpretations. Viewed as a tool for communication, the art map entails a quest for a new language to understand and experience place.

The Urban Space Rendered in Maps

Mapping always involves choices, requiring making decisions about what to map and what story to tell and defining the elements that will enhance and better communicate the spatial-temporal theme of the map, such as scale, symbology, colours, layout and design. Fundamental to these choices is the dilemma of what to include and what to leave out, which in mapping urban spaces is related to the definition of 'city' itself. Where should the boundaries of urban spaces be drawn? What are the aspects of 'urbanity' that should be depicted? What are the visual identifiers of a 'city'? These are some questions at the core of the problem of how to represent cities in graphic form, in maps that must be self-explanatory of their framework and purposes. To mapmakers the acts of 'depicting', 'describing' and 'narrating' urban spaces through maps involve a great deal of selecting images, data and stories, and making choices. The urban space imprinted on maps is a result of how it is perceived, by whom and in what conditions. As happens with any other form of expressing knowledge, maps are also vehicles to express biases and pre-conceptions that are historically constructed. It is therefore useful to explore here some of the narratives that shaped contemporary understandings of cities and investigate how they influenced the making of urban maps, starting with the concept of modernity.

One of the most enduring images of the contemporary city, that of a place of entrepreneurship that concentrates a diverse and multiskilled population, has its origins in the industrial era and the emergence of the capitalist metropolis and the culture of cosmopolitanism. The revolution in manufacturing processes initiated in Europe in the 1750s catapulted cities to the centre of a social, economic and cultural transformation in the nineteenth century, where railways, electric telegraphy and gas lighting became symbols of innovation.

[19] Jones, Tom (2012). Poetic language: theory and practice from the Renaissance to the present. Edinburgh: Edinburgh University Press, p. 1.

These transformations imposed a new perception of space and time and required a new concept to understand a world in constant change, and soon the mental images, depictions and discourses about the city were dominated by the theme of 'modernity'. The French poet Charles Baudelaire famously made the use of the word *modernité* to describe 'the ephemeral, the fugitive, the contingent' qualities that should be searched for the artist 'with an active imagination'.[20] Modernism became the label for an artistic movement that gave expression to the combined rational and emotional experiences of the new urban reality, embracing rationalist architecture and symbolist literature. Art forms emerging at the turn of the twentieth century embraced mechanical technologies, mass production and movement, and in no other place could all these novelties be experienced at once than in the city.

Walter Benjamin conceived modernity as an urban phenomenon and he situated the emergence of this modern world in 1820s Paris, taking the arcades—the covered shopping passages—as the epitome of a lifestyle founded in commodity capitalism.[21] The architecture of iron and glass became the international style of the industrial era, spreading throughout the main cities of the world such as Milan, Moscow, Johannesburg and Buenos Aires, and became associated with the splendour of the World Expositions or 'places of pilgrimage to the commodity fetish' according to Benjamin.[22]

In his 1903 essay *The metropolis and the mental life*, the German sociologist and philosopher Georg Simmel reflects upon the individuality and mental fragmentation of the 'metropolitan type' who, navigating an environment of sensorial overstimulation, 'reacts primarily in a rational manner' and supresses emotion in order to protect the 'inner life against the domination of the metropolis'.[23] This behaviour, characterised by experience a level of autonomy that functions both as a protective shield and a strategy that allows the individual to mix into the urban mass, resembles that of Benjamin's flâneur. However, while keeping the same 'blasé' attitude of Simmel's character, the flâneur in Benjamin's critique, which is drawn from

[20] Baudelaire, Charles and Mayne, Jonathan (trans.). (1995). *The painter of modern life and other essays.* 2nd edn. London: Phaidon, pp. 12–13.

[21] Benjamin, Walter (1999). *The Arcades Project.* Translation by Howard Eiland and Kevin McLaughlin. Cambridge, MA and London: Harvard University Press.

[22] Benjamin, Walter (1999). Op. cit., 'Grandville or the World Exhibitions', p. 7.

[23] Simmel, Georg. The metropolis and mental life. Levine, Donald (ed) (1971). *Simmel: On individuality and social forms.* Chicago: University Press.

literary depictions of Paris in the poems of Charles Baudelaire and the novels of Victor Hugo, acts more as an aloof observer of the city life.

The gaze of the flâneur, Benjamin wrote, betrays 'a profound alienation', deviating from the anxiety experienced by city dwellers and using the urban crowd as a veil through which 'the familiar city is transformed into phantasmagoria'. For Benjamin the abandonment of the flâneur to the enchanting atmosphere of the metropolis is revealing of the myth that sustains the fetishisation of the commodity, as 'in the person of the flâneur the intelligentsia becomes acquainted with the marketplace'.[24] Benjamin compared the flâneur to the gambler; the first attracted to the illusions of space, the second to the illusions of time.[25] The grim daily task of surviving, the dull corners of the city, the economic exclusion are all but hidden from the intoxicated gaze of the flâneur tempted by shop windows. The rag picker, 'the most provocative figure of human misery' according to Benjamin,[26] is the counterpart of the flâneur, collecting what is left behind, exploring the city on the fringes of capitalism.

Benjamin's counterparts, the flâneur and ragpicker, are archetypes of lonely figures that, engulfed by the ever-expanding modern metropolis, wander the streets in trajectories that are not predetermined but rather plotted as they move along. They depend on their ability to interpret the environment around them to be able to extract what they need, being the pleasure of the sight or to find whatever is useful for survival; in this sense their interaction with the city is performative. Nevertheless, throughout the twentieth century, it was the figure of the flâneur that remained the paradigmatic translator of the modern metropolis, associated with the culture of spectacle and the public life. The flâneur reinforced a tradition of thinking of cities as places to experience modernity and social interaction, where one can 'float' and absorb the energy and pace of human development. The flâneur also had long-lasting implications in artistic forms of mapmaking, influencing the shaping of the situationist drifter and the twenty-first-century psychogeographer,[27] as the bohemian nature of the flâneur would return to incite a more critical and adventurous exploration of the urban space, in the new context of super population and mega cities.

[24] Benjamin, Walter. (1999). Op. cit., 'Baudelaire or the streets of Paris', p. 21.

[25] According to Benjamin 'gambling converts time into narcotic'. Benjamin, Walter (1999). Op. cit., 'Haussmann and the barricades', p. 12.

[26] Benjamin, Walter (1999). Op. cit., 'Baudelaire', p. 364.

[27] As will be explored further in Chap. 4.

From the second half of the twentieth century, an unprecedented level of urbanisation was observed in every continent in the world, with particular attention to the emergence of large metropolitan areas in developing countries. That the world is now predominantly urban is a fact incorporated in governmental agencies' discourses and widely disseminated among the lay public. The United Nations (UN) was long engaged in predicting the crossing of the 50% threshold, when more than half of the world's population would inhabit urban settlements. In 2009 it was finally announced that the number of people living in urban areas had surpassed the number living in rural areas, and since then 'the world has become more urban than rural'.[28]

This global urbanisation brought into question a range of phenomena: the resurgence of shantytowns, the mega-metropolisation, the suburban sprawl, the gentrification of inner cities, the 'festivalisation' and rebranding of urban centres. The advent of the 'urban age', according to Brenner and Schmid, is the 'most influential contemporary meta-narrative of the global urban condition'.[29] However, this urban world is far from being homogenous and clearly defined, with no fixed boundaries, either cultural or physical, and cannot be characterised by general concepts. Some of the new mega-cities in Africa and Asia, such as Luanda and Bangalore, exhibit harsh contrasting landscapes of slums and skyscrapers. There are parts of the world exhibiting explosive processes of urbanisation, while others are stagnating and even disappearing. The inequalities of urban living are observed not only among developed and undeveloped countries but also within country regions and urban neighbourhoods, in a fractal-like condition, where parallel worlds of prosperity and precarity cohabit the same space at all scales.

Traditional categories of urban typologies, such as centre and periphery, are being dissolved in the configuration of megacities and metropolitan regions with supra-infrastructures obtruding the recognisable patterns of inner-city spaces, making it more difficult to separate the 'urban' from the 'rural'. These transformations are reflecting a destabilisation of the foundations of urban studies today, in the same way that in the late 1960s

[28] According to the United Nations, by the middle of 2009 the number of people living in urban areas worldwide reached 3.42 billion, in comparison with 3.41 billion living in rural areas (3.41 billion). See http://www.un.org/en/development/desa/population/publications/urbanization/urban-rural.shtml.

[29] Brenner, Neil and Schmid, Christian (2015). Towards a new epistemology of the urban? *City*, 19:2–3, 151–182; p. 155.

and early 1970s the epistemological crisis challenged 'positivist urban policy and quantitative urban geography'.[30] The problematic definition of 'cityness' persists as the myth of the 'urban age' is propagated in many variations that emphatically reproduce a popular discourse of urban supremacy. Neo-positivist paradigms applied to the knowledge of space—and particularly urban space—can be observed in the twenty-first century through ideologies that praise the 'science of cities' or the implementation of 'smart cities'. The 'smart city' concept involves the integration of physical infrastructures and services through systems like the Internet of Things (IoT), where computers, machines, objects and humans, through wearable technologies, can be identified and transfer data, collected by sensors attached to them, over a common informational network. Smart city technology can be used, for example, to monitor and manage computers and machines to enhance the efficiency of transportation systems and water supplies, or to improve public safety.

The 'smart city' discourse came forth in the mid-1990s, and over the subsequent decades it gained increased attention from the corporate world,[31] marketed as an unquestionable solution for urban problems. Governments from local to national levels fully embraced developments in ICT focusing on the benefits of making public services more efficient and facilitating urban management, pursuing the implementation of smart city projects in partnership with big corporations. In many parts of the world, the 'smart city' label has been applied to re-brand cities or new neighbourhoods, adding to the repertoire of prescriptions for utopian futures, such as 'creative cities', 'sustainable cities' and 'resilient cities', that have been incorporated into the contemporary language of urban planning and management.[32]

The initial enthusiasm with innovative forms of 'smart urbanism' was followed by criticism that pointed to the differences between discourse and practice and to the dangers of technocratic governance, deepening social and spatial discrimination or prioritising ICT networks over other

[30] Brenner, Neil and Schmid, Christian (2015). Op. cit., p. 154.

[31] A milestone in this process occurred in 2011, when the giant IT company IBM registered the term 'smarter cities' as its trademark, after applied to that two years earlier. Soderstrom, O. Paasche, T. and Klauser, F. (2014). Smart cities as corporate storytelling. *City*, 18:3, pp. 307–320.

[32] Soderstrom, O. Paasche, T. and Klauser, F. (2014). Op. cit., p. 307.

urgent needs in local governments' agendas.[33] As currently developed, the possible benefits and advantages of the smart city all-embracing utopia are still far from being a reality for the great majority of the urban population, as most often 'smartness' is limited to a few patches in the inner city, and it is out of the reach in the majority of urban areas.

Contrasts between the 'ordinary' and the 'smart' city can also be observed in cartographic terms. The imagery produced to represent the 'smart city' is optimistic and enthusiastic, depicting urban spaces that are clean of pollution, green and vibrant, a vision enriched by detailed maps that overlay intelligent networks onto the existent urban morphology. Ordinary urban spaces are poorly mapped, or not mapped at all. Around the world impoverished areas constitute 'blank spots' on city maps and remain absent from official programmes simply because the government lacks information about them. Only recently, in the past ten years or so, non-governmental or alternative initiatives were established to implement mapping and databases in such areas, which might not produce 'the glitzy pictures and techno-imaginaries of mainstream smart urbanism' but are attempting to tackle urban problems 'from the ground'.[34]

This shift to a more social inclusive mapping, bringing 'ordinary', less utopian and closer to reality narratives to the forefront of urban debates, is not entirely unprecedented. Throughout the history of cartography, urban mapping has been deployed to visualise new worlds, creating dream-like cities that instigate the imagination; it was to the same extent used to make sense of present realities, to record what is observed, without dismissing creative visualisations. In Britain two celebrated examples of 'scientific' maps that made original contributions to urban survey and mapping are those produced by the physician John Snow, illustrating the clusters of cholera outbreak in London in 1854, and the social researcher and philanthropist Charles Booth, demonstrating the levels of poverty in London (1889–1899).

These maps made use of cartographic tools in innovative ways to bring forth patterns that could not only inform but effectively contribute to scientific breakthroughs and decision-making.[35] In the case of Snow's map, his

[33] McFarlane, Colin and Soderstrom, Ola. (2017). On alternative smart cities. *City*, 21:3–4, pp. 312–328.

[34] McFarlane, Colin and Soderstrom, Ola. (2017). Op. cit., p. 313.

[35] Charles Booth's maps and notebooks can be explored in detail at the London School of Economics website https://booth.lse.ac.uk.

discovery, acknowledging him as one of the fathers of modern epidemiology, prompted to the removal of a water pump that he believed was the source of the cholera cases. Snow's map is simple and effective, using a symbology of stacked blocks that represent the number of fatalities and at the same time geographically pinpoint the residences of those affected by the illness. It is a clever and intuitive solution that makes the map easy to read. Booth's map is an advanced piece of cartography, exceptionally colourful and detailed, and his elaborated legend only enhances the visual language adopted to differentiate categories. Booth's mapping technique was widely used to inform public urban policies, and his survey methods influenced British sociologists as well as the Chicago School of Sociology in the USA.

While compelling examples of visual and semantic articulation, Snow's and Booth's are analytical maps that position the city as an object from which one extracts information to develop further urban planning strategies. They are produced to pursue scientific knowledge of the city and are incredibly powerful tools of communication, but in providing a useful and intuitive way to visualise spatial data, they have been used to assert certain ideas of the urban society. As cartographic outcomes of scientific enquiry, they must be understood within the historical, cultural and political context of urban theories and ideologies of the times of Victorian London, when the predominant theme was the urgency of 'healing' and 'cleansing' the city of its illness, both physical and social.

Towards the end of the nineteenth century survey maps were instrumental in justifying urban policies and programmes, becoming an integral part of modern urban planning, especially among Anglo-American planners. British cities, in particular London, were growing rapidly, and new suburbs were expanding into rural areas, which prompted the institution of regulatory bodies as well as the elaboration of pioneering works in planning theory and practice.[36] British thinkers were also among the most innovative authors to debate the state of urban planning and its methods, and it was down to a Scottish biologist, Patrick Geddes, to propose a revolutionary approach combining the use of surveying and mapping techniques.

Geddes' ideas were strongly influenced by late nineteenth-century French geographers and sociologists who emphasised an understanding of social, cultural and economic aspects of human life in relation to its

[36] For a more extensive history of urban planning in the UK and its impacts in the world, see Hall, Peter and Tewdwr-Jones, Mark (2011). *Urban and regional planning*. London: Routledge, 5th ed.

surrounding environment. From this, Geddes developed the concept of 'survey-before-plan', first presented in 1911 using the medical analogy of 'diagnosis before treatment', claiming that systematic surveys were conditional to successful urban planning.[37] Geddes' approach was multidisciplinary and holistic, considering cities as complex environments not apart from natural processes. Such ideas are at the foundation of overlay mapping technique, used to analyse sets of thematic maps at once, which in turn is the base of computing cartography as developed in geographic information systems.[38] The study and mapping of social and environmental interactions continued to interpolate urban and regional planning theories throughout the twentieth century, and the advancement of the use of computing statistical methods enabled the conception of evermore 'analytical' and 'accurate' maps of urban spaces.

Social theorists and cultural geographers have been critical of reductionist approaches to cities. The 'spatial turn' in humanities that emerged in the 1980s shifted attention to an understanding of urban space as a product of social action, therefore always mutable and subject to social dynamics.[39] According to this point of view, the city cannot be 'fixed' on its own by the wonders of pure science but transformed through social relations. This is not an entirely new concept: back in 1937 the American historian and sociologist Lewis Mumford, an admirer of Geddes' work,

[37] Batey, Peter. The history of planning methodology. In Hein, Carola (2011). *The Routledge Handbook of Planning History*. London and New York, Routledge. (pp. 46–59).

[38] The overlaying technique was first introduced in urban planning by British town planner Jacqueline Tyrwhitt in 1950, basing her work on Gedde's concept of 'survey-before-plan'. Tyrwhitt moved to the USA in 1955 to teach at Harvard Graduate Design School, while her overlay method became widely adopted in Britain. Later in 1967 American landscape architects Ian McHarg and Philip Lewis were the first to apply the technique using computer-generated maps and database. Hein, C. and Mil, Y. (2020). Mapping as gap-finder: Geddes, Tyrwhitt and the comparative spatial analysis of port city regions. *Urban Planning*, vol.5, issue 2, pp. 152–166.

[39] 'Spatial turn' is a term launched in the late 1980s by geographer Edward Soja and literary theorist Fredric Jameson, drawing on the works of French philosophers Michel Foucault (*Of other spaces*, lecture, 1967), Henri Lefebvre (*The Production of Space*, 1974) and their own analysis of the emergent 'postmodern condition'. The 'turn' in question is related to the shift from modern historicism to postmodern geographies, based on what geographer David Harvey called 'time-space compression'. Harvey, David (1990). *The condition of postmodernity; an enquiry of the origins of cultural change*. Cambridge, MA and Oxford: Blackwell. See also Soja, Edward (1989). *Postmodern Geographies: the reassertion of space in critical social theory*. London: Verso, and Jameson, Fredric (1991) *Postmodernism, or the cultural logic of late capitalism*. Durham, NC: Duke University Press.

advocated an interdisciplinary approach to planning that did not exclude its cultural and social dimensions. In his essay *What is a city?* Mumford argues that the city is 'a theatre of social action', where 'social drama' comes into existence. Urban planning, he added, can make it or frustrate it.[40] The 'spatial turn' critiques were also influenced by the writings of Henri Lefebvre and Michel de Certeau, which will be discussed in the next chapter, in directing urban theory to contemplate space as a social product and to consider the importance of everyday practices in conveying the city.

Humanistic views of the urban question in the late 1990s paved the way for new forms of activism and provided a counterweight to the concurrent 'neo-positivist' trend in spatial knowledge and information. Maps produced for planning and design are often visualised in realistic and dynamic 3D views that sustain a tension between utopian and dystopian landscapes. The assertion that they are maps based on reality—or possible ones—that has been studied, counted and analysed through scientific methods, does not imply that they are immune to the map makers' ideologies and standpoints. As Cosgrove argues:

> Indeed, the map may be the only medium through which contemporary urbanism can achieve visual coherence. [...] but today's acute awareness that cartographic images can never be innocent vehicles of information dissolves neat distinctions between celebratory and regulatory urban maps. Urban space and cartographic space remain inseparable; as each is transformed their relationship alters.[41]

In these alternative cartographic practices, often described as 'counter-mapping',[42] big data and geographical software applications are used as

[40] Lewis Mumford, *What is a City*, was first published in Architectural Record, 1937. For a more recent print, see Le Gates, Richard T. and Stout, Frederic (eds) (2011). *The City Reader*. London and New York: Routledge, pp. 91–95.

[41] Cosgrove, Denis. Carto-City. In Abrams, J. and Hall, P. (eds) (2006). *Else/where: Mapping New Cartographies of Networks and Territories*. Minneapolis: University of Minnesota Design Institute, pp. 148–165.

[42] The American anthropologist Nancy Lee Peluso was the first to use the term 'counter-mapping' in 1995 to describe mapping activism in the Indonesian forests. Since then 'counter-mapping' has been applied to any mapping activity envisaging the challenge and possibly disruption of power structures, providing alternatives to cartographic models enshrined by governmental agencies, urban planners and elite groups. Peluso, Nancy Lee (1995). Whose woods are these? Counter-mapping forest territories in Kalimantan, Indonesia. *Antipode*, 27:383–406. See also Culcasi, Karen. Community Mapping. In Monmonier, Mark (2015). *The history of cartography: cartography in the twentieth century* (Vol. VI). Chicago, London: The University of Chicago Press, pp. 286–287.

supporting tools to unveil and understand how human perception, reasoning and action transform cities and other human made environments. Not surprisingly artists engaged in this process of 'subverting' the dogmas of urban cartography, where the city is a subject to be criticised, reinvented and not necessarily be 'corrected' by the means of public policies or urban planning.

Contemporary Art and Mapping

In a talk given at the Drawing Center, New York, in 2004, the French philosopher Alain Badiou used a map as an example to support his theses on contemporary art. The map in question, *George W. Bush, Harken Energy and Jackson Stephens c. 1979–1990*, produced by the American conceptual artist Mark Lombardi in 1999, was an unusual piece of cartography, and it resembled a diagram more than a map. The large-scale drawing, as in a conventional map, indicated directions and connected nodes not of places but of individuals who had attained key roles in the (then) global network of power and politics. Badiou pointed out that Lombardi's map unveiled the real capacity of art 'to present something before the facts, before evidence' on creating 'a new possibility of art and a new vision of the world, our world'.[43] Badiou was then attacking the romantic view of art to emphasise his argument that contemporary art should seek to truly create *new* art. When he refers to Lombardi's work, it is to claim a *new* form of freedom that diverges from what he calls 'the imperial sense of freedom'.[44]

A common way to define what is contemporary is to refer to a chronological frame: if an artwork or art style is defined as 'contemporary', it refers to art that is produced 'today' or 'recently' or in 'our time'. Of course, this imposes a problem since time is not static and absolute, and all artworks were once contemporary. In whatever way we define limits of 'this time', as attempting a periodisation, the contemporary is still going to be arbitrary and relative, circumscribed by cultural and historical contexts, and in the case of 'contemporary art' such relativism brings on another problem: how to critically evaluate and define it as an entity in the history of art?

[43] See 'Fifteen Theses on Contemporary Art', *Lacanian Ink*, no. 23, 2004, pp. 103–119. Available at http://www.lacan.com/frameXXIII7.htm. Published article: Badiou, A. (2004). Thèses sur l'art contemporain, *Performance Research*, 9: 4, 86–86.

[44] Badiou, Alain. (2004). Op cit., p. 119.

Looking back at the twentieth century the American art critic and philosopher Arthur Danto considers 'contemporary art' a term that can be used to define a certain period in the history of art, designating 'something more than simply the art of the present moment'.[45] British philosopher Peter Osborne acknowledges that the common-sense idea of contemporary art as art produced at the present time, thus original, leads to the belief that such art production does not have a critical reference. Furthermore, he points out that part of this common-sense suggests that 'contemporary art' is a label for a specific type of present time art that is 'more living, more actual, and thus to be valued more highly than other art with which it, paradoxically, shares time'.[46]

Osborne notes that it is after the Second World War that the term acquires its current historical and critical connotations, becoming a credible label for artworks. Institutions exclusively dedicated to 'contemporary art' are first established in cosmopolitan cities in Europe and the USA and later elsewhere.[47] However, Danto observes that the distinction between 'modern' and 'contemporary' arts became clear only 'well into the seventies and eighties'.[48] In terms of avant-gardes Osborne identifies four phases[49]: the 'classical' 1910s' artistic and political movements, the 'neo' purely artistic of the 1940s–1950s, the revival in the 1960s–1970s and the challenge of the 'neo' in the 1980s. Danto, in the search of a definition that encompasses the postmodern and the contemporary, recognising at the same time that this is a period with no stylistic unity, suggests that contemporary art should be considered *post-historical art*. Osborne, seeking to make contemporary art 'critically intelligible' through philosophical discourse, proposed that contemporary art should be understood as *post-conceptual art*.

[45] Danto, Arthur C. (1997). *After the end of art: contemporary art and the pale of history*. Princeton: Princeton University Press, p.10.

[46] Osborne, Peter (2013). Anywhere or not at all: philosophy of contemporary art. London: Verso, p. 6.

[47] The first of these was the Institute for Contemporary Arts (ICA), founded in London in 1946, followed by the Boston Institute of Contemporary Art (1948), the City Gallery of Contemporary Art in Zagreb (1954), the Museu de Arte Contemporânea da Universidade de Sao Paulo (1963) and the Musée d'art contemporain de Montreal (1964), among others. Osborne, Peter (2013). Op. cit., pp. 16–17.

[48] Danto, Arthur C. (1997). Op. cit., p.11.

[49] Osborne acknowledges that this is partly based on Peter Burger's *Theory of the Avant-Garde* (1984).

Without going further in the philosophical journey into definitions and enquiries about the nature and purpose of art ('what is art?', 'what does art do?' and so on), what becomes evident in relation to contemporary art is the problematic issue of packing art works of a chronological period (which by itself is debatable) into one label. The way around it for Danto is to point out that the lack of a distinguishable style implies 'a period of information disorder [...] but equally a period of quite perfect freedom',[50] while Osborne talks of artworks being 'suspended' between the different avant-garde perspectives. Presumably the difficulty imposed by this undefinition is also the clue to distinguish artworks of the 'contemporary period' from their predecessors and artworks of 'contemporary style' from their synchronous counterparts.

Nevertheless, an important aspect of trying to define contemporary art considering its cultural and historical contexts is that related to the international hegemony of US art institutions after the Second World War mentioned by Osborne.[51] The American capitalist model was extended to the 'art world', as Danto defines it,[52] and as an affirmation of the power of the state, the public sector became a main sponsor of art galleries and museums. About that time a new market niche was created to promote and sell 'contemporary art', and over the last decades of the twentieth century, it became increasingly transnational and linked to global capital, attracting investors to art auctions and also to the promotion of large-scale international exhibitions. In the art map world such changes reflected in an increasing activity to promote exhibitions and publications oriented to maps as early as in the 1960s, accompanying the spread of 'contemporary art' in the international art world realm. It can be argued here that the distinction between 'art maps' and 'maps' is in part due to the pursuit of definitions that can be marketed in the art world of the Western world and

[50] Danto, Arthur C. (1997). Op. cit., p.12.

[51] Osborne, Peter (2013). Op. cit., p. 18.

[52] The term 'art world' is associated with the philosopher Arthur Danto, author of the influential essay entitled 'The Artworld', published in 1964. He proposed that a work of art is only identifiable in the context of artistic theory and history of art. Since then the term has been widely used to refer to a social and economic structure involving practitioners, critics, audiences, institutions and market. 'Art system', a less popular term, was proposed in 2001 by philosopher Larry Shiner to embrace 'the underlying concepts and ideals shared by various art worlds and by the culture at large'. See Danto, Arthur (1964). The Artworld. *Journal of Philosophy* 61, pp. 571–584; Shiner, Larry (2001). *The invention of art: a cultural history*. Chicago: University of Chicago Press, p.11.

Global North and that this distinction is associated with the structures for exhibiting and disseminating cartographic artwork established by this market.

Defining and contextualising contemporary art maps became a subject of scholarly study in the late 1980s, under the development of a critical thinking in geography dominated by Anglo-American academics such the American artist and professor Denis Wood and the British geographer Denis Cosgrove. Their work explored intersections between art and cartography trying to categorise the many and varied artistic manifestations to the generally defined 'art and science' of producing maps. They face similar problems as the philosophers trying to outline parameters for contemporary while acknowledging that definitions and categories obscure the nuances of multidisciplinary intersections, establishing boundaries that can become hard to trespass. Most of the history behind the evolution of contemporary mapping practices is better understood within the limits of the art movements that emerged in the 1910s and 1920s in Europe, namely, Surrealism and Dada, giving much of the conceptual foundations explored in art maps and mappings thereafter. Following Badiou's idea of the *new* we argue that during most of the twentieth century there was a string of new ideas, methods and philosophies that embraced the new against the traditional canons that shaped what we see in art maps today.

Considering contemporary artistic mapping practices, the attempts to better understand what goes on the art world cannot assume a general and simplistic definition of art maps—that is to say, maps produced by artists, not cartographers—as one characteristic observed is that the boundaries between the 'art and science' in cartography as in art are increasingly blurred. Artists are turning to the science of cartography, using up-to-date GIS and web mapping while cartographers and geographers are exploring art and collaborating with artists. The result is that this great diversity of practices enables one to interrogate the physical world and our relationship with it in more creative and critical ways. For this reason, there is an emphasis on the performative character of maps and mappings for purposes beyond *comment* over power and politics but rather *acting* as tools to change power and political stances.

The new practices of contemporary art mapping have been stimulating other discussion in arts, that is, the distinctions, or divisions, between art and craft, labour and creative work. Hannah Arendt in *The human condition*[53] deals with three aspects of human nature essential to define the circumstances of human existence: labour, work and action. *Labour* is related to routine that over time disengages body and mind, perpetuating a cyclical movement with no beginning or end. *Work* is related to fabrication, with a purposed beginning and end. *Action* concerns work that has definitive beginning but not a fully predictable end. Although all three are necessary to human life and survival, Arendt establishes a hierarchy that inevitably puts *action* in the category of the most thoughtful and sociable activity of human beings.[54] This point is made clearer in her description of *Animal laborans* and *Homo faber*: the first places the human being in the bestial condition to carry work without judgement; the second sees humans making things and elaborating over the making practice.

Half a century later Richard Sennett revised these arguments to analyse how labour and work are pertinent to craftsmanship, proposing a more 'balanced view' than that argued by Arendt, where 'thinking and feeling are contained within the process of making'.[55] Craftsmanship, implies Sennett, includes both manual thoughtless labour and insightful making, and one's set of skills are acquired by process of both repetition and innovation. Repetition involves 'bodily practices', the embodied knowledge that is 'gained in the hand through touch and movement', while innovation comes from the understanding of techniques 'through the powers of imagination'.[56]

In his book *The Invention of Art* the historian and philosopher Larry Shiner elaborates on Paul Oskar Kristeller's argument that what we understand today as 'fine arts' is a category established only in the eighteenth century; before that 'art' and 'craft' were indistinguishable, with artists

[53] Arendt argues that these categories belong to *vita activa* (active life) in opposition of *vita contemplativa* (contemplative life). The book was first published in 1958, and the events of the creation of the first atomic bomb in 1945 and the launch of the first space satellite in 1957 clearly impacted Arendt's ideas on the human capacity to challenge nature, at limited to the conditions of life. Arendt, Hannah (1958). *The human condition*. Chicago and London: The University of Chicago Press.

[54] 'All human activities are conditioned by the fact that men live together, but it is only action that cannot even be imagined outside the society of men. […] Action alone is the exclusive prerogative of man; neither a beast nor a god is capable of it, and only action is entirely dependent upon the constant presence of others'. Arendt, Op. cit. p. 22.

[55] Sennett, Richard (2008). *The craftsman*. New Have: Yale University Press, p. 7.

[56] Sennett, Richard (2008). Op. cit., p. 10.

also being considered artisans and vice-versa.[57] According to Shiner the modern system of separation between arts and crafts emerged in response to the development of a market among the middle classes. It is in this period that 'fine arts' translates into a series of dogmas such as 'art for art's sake', 'the artist as visionary genius' and 'unique aesthetics/artwork'. Art has no more function, or it transcends practical function (including decorative function): its only aim is to please the senses and to incite aesthetic experiences. This also coincides with the creation and expansion of new institutions devoted exclusively to the public appreciation of art, such as galleries, museums and concert halls.

This 'system of arts', as Shiner called it, of art categories and their respectively dedicated venues boosted a strong art market across Europe sustained by equally strong ideologies: in visual arts, for instance, the perfect representation ('academic' painting, portraiture) and the precious object (uniqueness, originality and rare materials, creating an expensive artwork). This 'traditional' art world could only firmly recognise fine arts as objects or performances that belonged to an artistic environment or network of credentialed artists, exhibited at designated art spaces and ratified by art curators and traders. It also emphasised the divide between the 'fine arts' (painting, sculpture, architecture, music and poetry) and 'popular' arts or entertainment.

These ideologies have endured down to the present, Shiner admits, but they were also affected and defied by processes of 'assimilation' and 'resistance': gradually assimilating new forms of art, from photography to jazz and graffiti, and at the same time resisting divisions that split art and society/life, 'attacking root polarities of the system itself'.[58] According to Shiner these processes started to become more evident in the early twentieth century, in movements such Dada, Russian constructivism and the Bauhaus, and since the 1960s, they accelerated. In relation to contemporary art maps a similar story can be told, as art maps and mapping became a recognisable category through the twentieth century, incorporating subversive practices, challenging art models or blurring the boundaries of the gallery and street/public space work. But this is far from being a progressive, continuous development. As Shiner pointed out, the old 'systems of arts' are still in place, noting that the art market is still a very strong

[57] Shiner, Larry (2001). *The invention of art: a cultural history*. Chicago: University of Chicago Press.

[58] Shiner, Larry (2001). Op. cit., p. 226.

institution that thrives on divisions and classifications, especially when they concern authorship, as to establish economic value to art works.[59]

However, a historic examination is helpful for the purpose of research and to elucidate the cultural context in which artists develop their mapping practices. In the geographical perspective of the Western world and Global North art world, it is possible to discern a periodisation that encompasses four phases. The first is an embryonic one, characterised by eventual artworks with maps of surrealist and dadaist artists; the second is represented by the work of the Situationist International in the 1950s; the third comprises experimental mapping and map artworks, associated with the conceptual art movement, that emerged in the 1960s and still has influence today; and the fourth phase starts around the mid-1990s and includes the consolidation of the art map as a genre, endorsed by art exhibitors and publishers, and the appropriation by artists of digital mapping technologies. This is a periodisation that will be discussed further in this book in Chap. 4.

Recalling Badiou's defence of the new in contemporary art, the following might be an example to illustrate the passage of time and changes that occurred in the appreciation and validation of mapping as an art practice. In 1979 the artist and educator Michael Daley wrote an article, published in Art Monthly (nowadays considered the UK's leading magazine in contemporary art), fiercely criticising the state of Fine Art education in Britain at that time. 'In many respects [...] graduating students today are inferior to their counterparts of 15 or 20 years ago', Daley affirms, suggesting that the teaching of standard skills, such as drawing, painting and sculpting, were neglected for the sake of following '*current* movements', to the point that students were allowed to leave art schools with portfolios 'full, not of drawings, paintings, etchings, etc. of some kind, but of such things as maps—perfectly ordinary maps—with lines on them indicating *walks*, allegedly undertaken by the graduating "artist"'.[60]

[59] This can be even observed when the art market assimilates the new and embraces forms of resistance: in October 2018, the graffiti *Girl with Balloon*, authored by the renowned street artist Banksy, was sold at auction for 1.04 million pounds. If the amount was not spectacular enough to make big news in the art world, the piece self-destructed just after the hammer came down. Specialists considered that this might increase the value of the artwork. See https://www.theguardian.com/artanddesign/2018/oct/06/banksy-sothebys-auction-prank-leaves-art-world-in-shreds-girl-with-balloon.

[60] Daley, Michael (1979). The debilitation of the British art school. *Art Monthly*, 7, pp. 29–31. (author's emphasis in italic).

Regardless of Daley's personal views on what should be considered 'fine art' or not, it is interesting to point out that mapping was a subject in art students' works at that time and in a performative form, as indicated by Daley's mention of 'walks'. It would be interesting to further investigate if such deviations were a trend or just Daley's unfortunate single encounter with a group of young 'map artists', but with the impossibility of doing so, it is worth noticing that his critique coincides with the early years of the acceptance of land artists, including the acclaimed Richard Long, into the mainstream art world and its prestigious art galleries.[61] Most probably the students recreating their own mapping works were influenced by what they were seeing in these art spaces, in which case they would be more attuned with the new and contemporary art world than their lecturer would suppose.

Having explored the key concepts of 'translation', 'urban' and 'contemporary art', the next chapter will bring discussions on further theoretical approaches to the meanings of 'seeing', 'representing' and 'performing' in the perception and understanding of place and the urban space and how maps and mapmaking are interconnected with knowledge and reality.

[61] In 1973 Tate Gallery acquired Richard Long's *A Hundred Mile Walk* (1971), and in 1976, another two of Long's walking pieces: *A Line made by walking* (1967) and *Cerne Abbas Walk* (1975).

CHAPTER 3

Seeing, Representing, Performing

A map is basically a projection of the physical world that gives an interpretation of reality. What does mapmaking entail? In principle, the fundamental work of mapping is observing and registering a landscape. However, these simple actions conceal a complexity of decisions, from what exactly is observed to how to proceed in registering it and what methods to employ in each stage. Then, the discussion continues: as the object produced is a result of choices, technical limitations and subjective presumptions, how can it be described as a 'true' representation? That seems easy to answer: no map is completely true. As the scientist Alfred Korzybski famously stated, 'a map is not the territory',[1] but the conceptual muddle of what separates reality from representation persists throughout the history of map making, and in an era of virtual reality, such debate is unavoidable.

A more interesting question could reside in not identifying the dissimilarities between one and the other, the thing and its symbol (or using the concepts of semiology, the sign and the signifier) but how territory and

[1] Originally the phrase first appeared in an academic paper of 1931, reprinted later in the book *Science and Sanity* (first edition, 1933). The full paragraph reads like: 'A map *is not* the territory it represents, but if correct, it has a similar structure to the territory, which accounts for its usefulness. If the map could be ideally correct, it would include, in a reduced scale, the map of the map; the map of the map, of the map; and so on, endlessly' (emphasis in the original). Korzybski, A. (1948). *Science and Sanity: An Introduction to Non-Aristotelian Systems and General Semantics.* 3rd ed. Lakeville: The Internal Non-Aristotelian Library, p. 58.

G. Lanci, *Art Maps and Cities*,
https://doi.org/10.1007/978-3-031-13306-0_3

map are entangled in the human endeavour of communicating perceptions and ideas about place and space. In the following sections I present some debates around how maps work, and consequently what mapmaking entails, organised around the concepts of seeing, representing and performing. The first section of the chapter elaborates on the dialectical relationship between the view 'from above' and the view 'from the ground' to present phenomenological approaches to urban space and theoretical debates about the cultural construction of acts of seeing and how that affects the perception of space. The second section discusses modes of representing the world and their implication for spatial knowledge, how critical thinking contests cartographic traditions of representations and alternative propositions of performance, agency and deep mapping.

The View from Above, the View from the Ground

On December 24, 1968, the dream of seeing the entire Earth from above, a fantasy nurtured for centuries in human history, depicted in myths, narratives, legends and images, was turned into reality. On that day the three astronauts of the Apollo 8 mission took several pictures in colour film of the Earth surfacing behind the lunar orbit. One of them became the iconic *Earthrise*, the popular name given to image AS08-14-2383. The photograph of the Earth rising from behind the Moon was charged with symbolic power: emulating the clichéd picture frame of a sunrise in the horizon, it transcended what one could imagine as a view from outer space. This was the ultimate 'God's eye view', a poetic and sublime picture of our planet. The strength of this correlation, between the raw accurate image of the planet taken by the achievements of human technology and the idea of looking through the lens of an ethereal supreme being, appealed to people, and it was perpetuated through a variety of media, for example, the stamp issued by the US government in 1969.

The space technology of the second half of the twentieth century that engendered satellite imagery and remote sensing[2] also provided a way of seeing that was long pursued in cartographic representation: objective, precise and, above all, realistic. This came to reinforce the surveying tradition in cartography and the idea of the map as a product of accurate visual

[2] The first orbital photographs of Earth were made in 1959 by the US satellite Explorer 6. The Landsat programme, aiming to comprehensive acquisition of imagery of Earth from space, was launched in 1972, also by the American government.

representation of the land. Yet the act of seeing is not divorced from perception, and as the case of the Apollo 8 imagery exemplifies, it involves a mental process that selects points of view and angles of observation. Representations of the globe, as popularised by the Apollo 8 and subsequent moon missions in the early 1970s, have other implicit meanings though. Culturally they perpetuate colonial and imperial perceptions of the world, which have been constructed upon the European early modern rhetoric of the fifteenth and sixteenth centuries, when the first world atlases were published. The reproduction and dissemination of the Apollo space images bear similarities with ideologies prevalent back then, when cartographers represented European global knowledge in maps, joined with 'universalizing Christian cosmology and planetary geo-strategic discourses'.[3]

The same approach applies to the early modern representations of cities: viewed from above they were featured as universal truths, depicted on maps capable of shrinking the cityscape enough to allow the illusion of holistic knowledge that potentially empowers the viewer. Apprehending a reality much larger than us in one glimpse instigates a sense of wonder and at the same time domination. It also reduces the complexity of its structure as the city can be therefore tamed and smoothed of its irregularities, discrepancies, imperfections and roughness. The view from the ground implies being deliberately partisan and partial, selecting what to depict on the map based on personal choices: it is an intuitive approach. From above the view tends to be uniform, impartial, depicting 'facts' that can be comparable, measurable, categorised: it is a rational approach intentionally framed as 'neutral'.

Aerial photography, before the advent of satellite imagery, was the technology that grabbed the imagination of artists and city planners alike. The earliest record of urban photographs taken from above is credited to the French photographer, journalist and enthusiastic balloonist Gaspard-Felix Tournachon, more popularly known as Nadar, who first flew over Paris in

[3] Cosgrove, Denis (1994). Contested global visions: one-world, whole-Earth and the Apollo space photographs. *Annals of the Association of American Geographers*, vol. 84, 2, pp. 270–294. Further comments and analysis can be found in Cosgrove, Denis (2001). *Apollo's eye: a cartographic genealogy of the Earth in the Western imagination*. Baltimore and London: John Hopkins University Press; and Poole, Robert (2008). *Earthrise: how man first saw the Earth*. New Haven and London: Yale University Press.

1858. After him, other techniques were employed,[4] but it was during the First World War that aerial photography was greatly developed, including purpose-built cameras and aeroplanes. In 1919 Gordon Holt, an RAF pilot and photographer, published an article in the *Architectural Review* arguing for the usefulness of aerial photography in architecture, explaining how this (then) new technology was 'bound to affect the method of obtaining the plan or perspective of an actual building, street, or of a whole town'. Suggestively, Holt uses the eye as a metaphor for the efficacy of the system:

> Now, in aerial photography, the eye is not assumed to be there. *It is there*, and from its point of vantage it – or rather the lens – merely records everything occurring in this given area, in its three dimensions of length, breadth and depth.[5]

The aerial view observed by human eyes, taken from hot air balloons or airplanes, and transformed into corresponding images through photographic techniques, had a great impact in artistic forms at the onset of the twentieth century. The aerial view of cities impregnated works from early modernist art movements: cubists, futurists and abstractionists embraced the creative impulse offered by those images,[6] a testimony to the cultural and technological revolution of their time, inscribed in the accelerated development of the urban landscape. Kazimir Malevich, the founder of the Suprematism art movement, was one of the artists inspired by the aerial view back in the 1910s, in developing his own style of abstract paintings.[7] The detachment that the aerial view implies was explored in Malevich's paintings as transformed realities, made possible with a new perception of urban landscapes from above.[8] In the years after the Second World War the technology developed to survey (as well as espy) the territory from above left a legacy of aerial imagery that was utilised by architects and urban

[4] Kite aerial photograph was pioneered by British meteorologist E. D. Archibald in 1882 and also used in France by Arthur Batut in 1888.

[5] Cited in Macarthur, John. On the obliqueness of the plan in urbanism and architecture. Dorrian, Mark and Pousin, Frederic (eds.) (2013). *Seeing from above: a cultural history of the aerial view*. London: I.B. Tauris, pp. 188–209; p. 189.

[6] Roseau, Nathalie. The city seen from the aeroplane: distorted reflections and urban futures. Dorrian, M. and Pousin, F. (eds.) (2013). Op. cit., pp. 212–213.

[7] Lodder, Christina (2013). Transfiguring reality: Suprematism and the aerial view. Dorrian, M. and Pousin, F. (eds.) Op. cit., pp. 95–117.

[8] Lodder, Christina (2013). Op. cit., pp. 101–102.

planners in the reconstruction of European cities. Artistic interpretations of survey maps supported a torrent of Cold War propaganda for mass communication, identifying on maps all the enemy nations and all possible routes for the enemy's invasion. Attached to the Cold War came the Space Race, and then pictures from space—like the Earthrise—gained new symbolic meanings.

Modes of viewing the world mediated by technical apparatus instigated the question of how much our perception and sense of reality is dependent not only on *what* is seen but on *how* it is seen and therefore to examine the role of subjectivity in achieving knowledge. In the first decades of the twentieth century, a new philosophical movement was founded based on the study of phenomena as they are experienced by human consciousness, appropriately called 'phenomenology'. As a school of thought, phenomenology was first developed by the German philosopher Edmund Husserl in the 1920s, proposing a new philosophical method that considers the phenomena[9] of immediate experience and direct intuition as means of knowing reality. Phenomenology, as described by Merleau-Ponty, one of Husserl's most influential followers, is a philosophy that is not about reflecting upon 'pre-existing truth but, like art, [it is about] the act of bringing truth into being', because 'true philosophy consists in relearning to look at the world'.[10]

Many scholars after Husserl resorted to phenomenology and its variants,[11] and combined with poststructuralism and postmodernism philosophical critiques of positivism and Cartesian explanatory models, they developed theories about place and space with a focus on human's perceptions and sensorial experiences, approaching the physical environment, natural or human-made, not as an abstract entity but as a space of lived existence. Two of these theoretical developments further explored the nature of the urban space and its representation: that of the French social

[9] 'Phenomenon' comes from Greek *phainomenon* 'that which appears or is seen', noun of the verb *phainesthai* 'to appear'. Phenomenology thus reflects upon the way the world appears or reveals itself to human beings.

[10] Merleau-Ponty, Maurice (2002). *Phenomenology of perception*. London: Routledge Classics, pp. xxii–xxiii.

[11] Phenomenology's commitment to describe things as they appear meant a freedom to engage with diverse disciplines, and so, along the twentieth century, it generated diverse theoretical projects. Renowned phenomenologists include Gaston Bachelard, Merleau-Ponty, Martin Heidegger, Jean-Paul Sartre, Hannah Arendt, Emmanuel Levinas and Jacques Derrida.

theorists and philosophers Michel de Certeau and Henri Lefebvre. Although regarding their political and philosophical stands they had little in common—Certeau was an anti-Marxist Jesuit scholar while Lefebvre was deeply engaged with Marxist theory and its critique—they were both contemporaries of the situationists and experienced the events in Paris in 1968 (Certeau as an observer; Lefebvre as an active participant), and their works are often cited together, especially Certeau's *The Practice of Everyday Life* (*L'invention du quotidien*, 1980) and Lefebvre's *The Production of Space* (*La production de l'espace*, 1974).[12] Translated into English in 1984 and 1991, respectively, these works have found a receptive audience among Anglo-American scholars in the fields of urban studies, cultural studies and cultural geography, as well as having had reflections in the consolidation of critical cartography.

The Dialectics of the Lived Space

In an often-quoted paragraph from *The Practice of Everyday Life*, Michel de Certeau describes his experience looking at New York from the top of the World Trade Center as undergoing a metamorphosis of the senses that radically changes the perceptions of the viewer:

> To be lifted to the summit of the World Trade Center is to be lifted out of the city's grasp. One's body is no longer clasped by the streets [...] His elevation transfigures him into a voyeur. It put him at a distance. It transforms the bewitching world by which one was 'possessed' into a text that lies before one's eyes. It allows one to read it, to be a solar Eye, looking down like a god.[13]

The aerial view, Certeau implies here, is equivalent to that of a supreme being who apprehends the city as a whole, empowered by the elevated position 'above' that reality. Following up these reflexions, Certeau contrasts this granted power, from the advantage point of a skyscraper, with

[12] Certeau and Lefebvre wrote prolifically and widely, but for the purpose of discussing space and spatiality, I will concentrate in these two major works while citing other books. Interesting to note is that Certeau's English version of *The Practice of Everyday Life* was available much before Lefebvre's *The Production of Space*, but it was written six years after Lefebvre's book.

[13] Certeau, Michel de (1984). *The practice of everyday life*. Translated by Steven F. Rendall. Berkeley: University of California Press, p. 92.

the view from the ground, 'down below', where the practitioners of the city live. He then questions what can be really seen from above: is it the city known and lived by its inhabitants or a 'representation, an optical artefact'? Certeau argues that this panoramic cityscape seeing from above is a visual simulacrum that eludes the observer—here in the position of a 'voyeur'—and alienates the ordinary city life.[14]

The categories of 'above' and 'ground' in Certeau's work are associated with the perspectives of 'voyeurs' and 'pedestrians' over the city. The voyeur's experience of looking at the urban landscape with the 'god's eye' replicates in the same way the perspective view enacted in Medieval and Renaissance paintings; Certeau points out: the tower over Manhattan perpetuates the fiction attempted in those paintings that is possible to make the city legible at once. The city, for the voyeur, is packed into a concept. However, Certeau argues that the life of the city cannot be contained in the concept of the city, which reduces the 'urban fact', the bodily experienced city, to a rational or 'operational' entity for administrative purposes:

> Linking the city to the concept never makes them identical, but it plays on their progressive symbiosis: to plan a city is both to think the very plurality of the real and to make that way of thinking the plural effective; it is to know how to articulate it and be able to do it.[15]

In the conceptual discourses of the urban, Certeau observes, space itself risks being forgotten, becoming 'the blind spot in a scientific and political technology'. Thus, for him spatial practices 'from the ground' and the everyday are to counter-balance totalising views of the city exercised through the 'god's eyes' of the city manager, the urban planner or the cartographer. The urban lived space springs up on everyday practices, like walking, describing, narrating and remembering the city, which are analysed in detail by Certeau in his book as a phenomenology of embodiment.

Certeau's theorisation of space is much connected with the phenomenological conception of space of Merleau-Ponty, and it draws on ideas of spatial perception and categories of 'geometric space' and 'anthropological space'. Certeau makes direct reference to these concepts on calling for 'another spatiality' (another term appropriated from Merleau-Ponty) to locate the practices in a 'anthropological, poetic and mythic experience of

[14] Certeau, Michel de (1984). Op. cit., pp. 92–93.
[15] Certeau, Michel de (1984). Op. cit., p. 94.

space',[16] which takes the body at the centre of perception of the space and the expression of that perception. However, in Certeau's work there is a clear distinction between 'place' and 'space', used to emphasise two different approaches to the physical world. A 'place', he argues, is indicative of the order or location of elements, and it is a stable configuration of their positions, while a 'space' exists when other variables are considered, such as direction, speed and time, and it is a mobile composition of such variables. As he concludes:

> In short, *space is a practiced place*. Thus the street geometrically defined by urban planning is transformed into a space by walkers. In the same way, an act of reading is the space produced by the practice of a particular place: a written text, i.e., a place constituted by a system of signs.[17]

This theoretical reasoning adds a temporal dimension to the subject, avoiding an opposition between categories (as in the geometric/anthropological categories) and contemplating the fluidity of such definitions. Certeau suggests that 'place' and 'space' are determined historically, not in a chronological sense but as a string of stories in everyday practices that 'constantly transforms places into spaces or spaces into places'.[18] This idea is exemplified by the analysis of oral descriptions of places according to the categories 'tours' and 'maps' as presented in a study conducted in New York:[19] the descriptions related to the 'tour' type are those that provide instructions to act or move through space (or between places), while the 'map' type are descriptions that indicate the spatial order or location of the places.

Certeau observes that the itinerary is 'a discursive series of operations', related to 'acting', and the map is a 'place projection totalizing observations' and is related to 'seeing'; one implies a local order and the other produces a representation of places, and both are 'symbolic and anthropological languages of space'.[20] He makes clear that these are 'predominant' types of descriptions in a discourse, which does not need to be restricted to one type or another. Furthermore, he points out that maps were not

[16] Certeau, Michel de (1984). Op. cit., p. 93.
[17] Certeau, Michel de (1984). Op. cit., p. 117 (italicised in the original).
[18] Certeau, Michel de (1984). Op. cit., p. 118
[19] Linde, Charlotte and Labov, William (1975). Spatial Networks as a Site for the Study of Language and Thought. *Language*, 51 924–939. Apud Certeau, Op. cit., p. 119.
[20] Certeau, Michel de (1984). Op. cit., p. 119.

always a rigorous representation of places, and it was only from the fifteenth century onwards that they started to be shaped into the geographical form that is known today, leaving behind its pictorial narratives and becoming a totalising base to organise spatial knowledge.[21] This critical perspective in Certeau's work, reinforcing the importance of human agency while resisting to bureaucratic concepts of space, has been related to revolutionary discourses of the situationists in the 1960s and Lefebvre's *Critique of Everyday Life*[22]; however, it was pointed out that Certeau did not take the point of view of his predecessors about changing the world and going against mass culture, but rather he stressed 'the emancipatory and creative nature of consumption and of popular culture'.[23]

Lefebvre thought of space as developed in a dynamic, dialectical way, considering the complex interactions of human activities as transformative agents. In *The Production of Space* he looks for a 'unitary theory' of space as a social practice, where behaviours and actions upon the physical world are intrinsically connected to its production. The main concept implies that space is not a mere container for human activity; it is instead actively 'produced' through human activity, which in turn transforms space. This is clearer to observe in cities, where the dynamics of production are accelerated by the concentration of individuals constantly interacting with each other and with the environment around them; their behaviour is being shaped by the physical structure while acting to change it.

Considering space as a social phenomenon, Lefebvre emphasises the importance of the human body to understand how space is constituted through means of power. He identifies strata of spaces, proposing a dialectical triad of spaces: *spatial practice*, *representations of space* and *representational spaces*. The first category, *spatial practice*, is related to the actions in the physical space, a paradoxical and 'close association, within perceived space, between daily reality (daily routine) and urban reality (the routes and networks which link up the places set aside for work, 'private' life and leisure)'.[24] This is the material space of experience and perception. The

[21] Marin makes a similar observation, which will be discussed in the next section. Marin, Louis (1993). Frontier of Utopia: past and present. *Critical Inquiry*, vol. 19, n. 3 (spring) pp. 397–420.

[22] Lefebvre initiated a series of books entitled *Critique de la vie quotidienne* as early as 1947, followed by *Critique de la vie quotidienne II* in 1961 and *Critique de la vie quotidienne III* in 1981.

[23] Gardiner, Michael (2000). *Critiques of everyday life*. London: Routledge, p. 161.

[24] Lefebvre, Henri (1991). *The production of space*. Oxford: Blackwell, p. 38.

second, *representations of space*, is tied in with the modes of production through knowledge, codes and signs. This is the conceptualised space, 'the space of scientists, planners, urbanists, technocratic sub dividers and social engineers, as of a certain type of artist with a scientific bent—all of whom identify what is lived and what is perceived with what is conceived'.[25] The third category, *representational spaces*, embodies complex symbolisms and is 'linked to the clandestine or underground side of social life, as also to art'. This is the 'experienced space', tending towards 'more or less coherent systems of non-verbal symbols and signs'[26] (whereas the representations of space are coded in textual forms). This is the space of sensations and imagination incorporated into daily life.[27]

The perceived (*perçu*), the conceived (*conçu*) and the lived (*vécu*) (respectively, spatial practice, representations of space, representational spaces) aim to divert from 'oppositions, contrasts or antagonisms' and to go beyond the dualism and binary theories that 'strips everything which makes for living activity from life, thought and society', obscuring knowledge.[28] Lefebvre explains his triad with an analogy of the body: the physical body that senses the real world through its organs belongs to the realm of the *perceived*, the body that is examined and described according to scientific knowledge is derived from *representations* (of the body) and the body that experiences the world through culture, symbolism, creativity, is the body *lived*.

The lived city is understood as a space of friction and collective interaction that are constantly moving and changing; in this sense representations of space can never be totally true to the endless diversity of urban spaces. When exposing his idea of social space that is what Lefebvre observed in relation to maps, as he points out that no finite number of maps can deal with a given space:

> What we are most likely confronted with here is a sort of instant infinity [...] It is not only the codes – the map's legend, the conventional signs on map-making and map-reading – that are liable to change, but also the objects represented, the lens through which they are viewed, and the scale used [...]

[25] Lefebvre, Henri (1991). Op. cit., p. 38.

[26] Lefebvre, Henri (1991). Op. cit., p. 39.

[27] It can be observed here that the definition of *representations of space* is comparable to Certeau's notions of *place* and the 'concept-city', while *representational spaces* is closely related to Certeau's idea of *space*.

[28] Lefebvre, Henri (1991) Op. cit., p. 39.

> We are confronted not by one social space but by many – indeed, by an unlimited multiplicity or unaccountable set of social spaces.[29]

The 'conventional signs' of the map, in the 'descriptive or geographical sense', as Lefebvre puts it, cannot represent reality, because reality is mutable and multi-spatial. Classification and categorisation are useful strategies to simplify the world and cope with its complexity, but they risk homogenising space. This is a critique that later, in the mid-1980s, resonated among geographers who attacked the dual thinking and the Cartesian approaches to science, which do not recognise haziness and indistinct borderlines.

Lefebvre elaborates his critique of orthodox theories (including orthodox Marxism) in order to demonstrate how the constraints and regulations that dominant forces impose upon the urban space alienate humans from the lived practices in the city. He affirms, however, that the three instances of space should be interconnected and 'generalized in their application to cover all societies, all periods, all 'modes of production',[30] reiterating his objective to develop a unitary theory, where space is simultaneously perceived, conceived and lived. He points out that throughout history societies went from periods of the representations of space to representational spaces, which are also detected in works of art, citing examples in painting, literature and architecture.[31]

Lefebvre delineates a particular history of space that does not follow a chronological chain of historical events but unfolds according to the development of productive process, where he identifies three spatial types: ***absolute space***, ***abstract space*** and ***differential space***. He relates ***absolute space*** to those natural sites appropriated by humans for religious and political purposes, transforming the land into settlements and places of worship, and the ***abstract space*** is that of capitalist accumulation, where productive forces are alienated from the means of production and consequently from the land and nature.[32] These two types are well defined in

[29] Lefebvre, Henri (1991) Op. cit., p. 85.

[30] Lefebvre, Henri (1991) Op. cit., p. 41.

[31] Related to architecture he cites Frank Lloyd Wright, who 'endorsed a communitarian representational space deriving from a biblical and Protestant tradition', and Le Corbusier, who 'was working towards a technicist, scientific and intellectualized representation of space'. Lefebvre, Henri (1991) Op. cit., p. 43.

[32] This historical overview of the conception of space is detailed in the second half of *The Production of Space*, corresponding to Chap. 4 'From absolute space to abstract space', Chap. 5 'Contradictory space' and Chap. 6 'From the contradictions of space to differential space'.

Lefebvre's argument, but *differential space* falls into a nearly utopian category, as it is considered a reaction to abstract space and its conflicts:

> [...] abstract space harbours specific contradictions. [...] the reproduction of the social relations of production within this space inevitably obeys two tendencies: the dissolution of old relations on the one hand and the generations of new relations on the other. Thus, despite – or rather because of – its negativity, abstract space carries within itself the seeds of a new kind of space. I shall call that new space 'differential space'.[33]

Lefebvre envisages that abstract space, characterised by the homogenised city produced by bureaucrats and technocratic design, can generate a differential space where the plurality and diversity of the everyday life thrive, relieved of the oppressive forces that control the city. In previous works published just before and after the Parisian revolts of May 1968,[34] Lefebvre started to elaborate his critique of urban development under capitalism and the need to reclaim the city.

The concept of differential space echoes ideas formerly advocated by Lefebvre that in *The Production of Space* are reinforced by a call to address—or confront—the contradictions of abstract space through a collective appropriation of the space, as the 'right to be different' is only meaningful when derived from 'actual struggles to establish differences'.[35] Concluding the book Lefebvre reaffirms what he considers a project of 'a different society, a different mode of production, where social practice that would be governed by different conceptual determinations'[36] is intertwined with the understanding of social production as a key element of spatial dynamics:

> The transformation of society presupposes a collective ownership and management of space founded on the permanent participation of the 'interested parties', with their multiple, varied and even contradictory interests. It thus also presupposes confrontation - and indeed this has already emerged in the problems of the 'environment' [...] On the horizon, then, at the furthest

[33] Lefebvre, Henri (1991) Op. cit., p. 52.

[34] Le droit a ville (1968) was published before May 1968, followed by L'irruption de Naterre au sommet (1968), La revolution urbaine (1970) and La pensée marxiste et la ville (1972).

[35] Lefebvre, Henri (1991) Op. cit., p. 64.

[36] Lefebvre, Henri (1991) Op. cit., p. 419.

edge of the possible, it is a matter of producing the space of the human species […][37]

There is in those lines a manifestation of the spirit of change and rebellion of an intense period in Lefebvre's public and personal life, from leaving the French Communist Party and joining the situationists in 1958, until the fallout with Guy Debord and the events of May 1968. His theories on urbanisation and the production of space are in part a result of that experience, which in turn engendered other ramifications in the arts and mapping practices influenced by the Situationist International.

Before examining these more specific repercussions of Lefebvre's work, and in particular the contribution of the Situationist International in Chap. 4, the second part of this section will present some theoretical debates about acts of seeing and the perception of space.

Theories of Perception, Ways of Seeing

In many cultures sight has been long considered one of the most valuable of the human senses, enabling an immediate connection to the world, an instant perception of reality. What is seen becomes truth and what is hidden is unacknowledged; 'out of sight, out of mind' is the saying. Greek philosophers who theorised over vision argued that to see an object was to touch the soul and that nature can be only known through one's eyes: to know is to see.[38] The eye has a central part to play in this logic and serves as trope to various forms of apprehending the world, either being the 'interior eye' of religious metaphors, the 'exterior eye' that empirically examines nature or the 'mediated eye' transmuted by technology.

The language of Western philosophical tradition was developed upon visual metaphors that enhanced an 'ocular-centric' culture dominated by sight. However, as much as sight could be related to 'enlightenment' and 'knowledge', it was also acknowledged as leading to deception and falsity, for as the human vision can be tricked, so the mind. Reactions to cultures of vision were manifested in religious thought since the Middle Ages, and the development of optics in the seventeenth century proved that the connection between the eye and the soul was unfounded, but it was only in

[37] Lefebvre, Henri (1991) Op. cit., p. 422.

[38] Pomian, Krzystof. Vision and Cognition. In Jones, Caroline A.; Galison, Peter (eds.) (1998). *Picturing Science, Producing Art.* New York: Routlegde, pp. 212–213.

the early twentieth century that an *anti-visual* or *anti-ocular* discourse was structured in philosophy as in arts, interrogating the culture of visual supremacy.[39] John Berger in his most acclaimed book *Ways of Seeing*[40] drew attention to the fact that our gaze is culturally produced and impregnated with ideologically filtered views of the world. Class and gender are highlighted in having a major effect in how sight and mind are connected in Western society, as well as imperialism, which often goes in tandem with technological and race biases to exacerbate power and dominance over other cultures.

The perception and cognition of the external world came to a dramatic change in the Renaissance period with the discovery of linear perspective by architect Filippo Brunelleschi, and its application for rendering the observed three-dimensional space into the flat surface of the canvas redefined European painting in the fifteenth and sixteenth centuries. Art historian Erwin Panofsky, in his 1927 essay *Perspective as symbolic form*,[41] argued that beyond its technical achievement the geometrical mathematics-based perspective that emerged in the Renaissance was rather the expression of a 'worldview' settled in a particular time in history, when a cosmological approach of the universe (infinite and not centred, as suggested by Blaise Pascal[42]) was predominant. Panofsky points out that the question regarding artworks of that period is not whether they are constructed according to the correct rules of perspective but what kind of perspective they present,[43] meaning that each artwork has its own perspective that reflects a specific worldview in the larger cultural context.

In his treatise *Della Pittura* (*On Painting*), published in 1450, artist and polymath Leon Battista Alberti famously wrote *Dove a me paia, fermo uno punto*, 'I fix a point wherever I like'.[44] This arbitrary point, from which

[39] Jay, Martin (1994). Downcast eyes: the denigration of vision in twentieth century French thought. London: University of California Press.

[40] *Ways of Seeing* is based on the homonymous BBC television series, broadcasted in 1972. The book was chiefly written by Berger in collaboration with Michael Dibb (TV series producer), Sven Blomberg, Chris Fox and Richard Hollis. Berger, John (1973). *Ways of seeing*. London: British Broadcasting Corporation and Penguin Books.

[41] Panofsky, Erwin (1991). *Perspective as symbolic form*. New York: Zone.

[42] Pascal refers to nature as 'an infinite sphere, the centre of which is everywhere, the circumference nowhere'. Pascal, Blaise. (2016) *Pensées*. New York: Philosophical Library, Section 2, 72.

[43] Panofsky, Erwin (1991). Op. cit., p. 41.

[44] Spencer, J.R. (1966). *Leon Battista Alberti on painting*. Revised edition, Yale University Press, Book One, p. 57.

perspective is generated, endorses the subjectivity of the observer (artist-cum-cartographer) and consequently the particularity of the work produced. To select one point is to choose one's place in the world, and this viewpoint affects what we see, not only in empirical terms (the place where one stands, physically, and looks over the landscape) but also in metaphorical terms (the place in history and society one stands within). Therefore, while seeing the same landscape, a group of people, placed physically in the same spot, might not actually agree what meanings, emotions and thoughts that landscape evokes. As David Harvey noticed, perspectivism implied seeing the world from the standpoint of the individual who represents 'what he or she sees as in some sense "truthful", compared to superimposed truths of mythology or religion'.[45] The Renaissance painter is an observer, a spectator of the world, emotionally withdrawn from the scene in order to depict it faithfully through geometric rules. The visible space is ordered and duplicated in exactitude, and it is in the eye of the beholder that the world is organised—a humanist eye, not a godly one.

The perspectival painting presents a 'uniform, infinite, isotropic space that differentiated the dominant modern world view from its various predecessors'.[46] This emergent modern world was embracing the scientific understanding of nature and the reorganisation of the economy based on a capitalist system, which would be later implemented in the conquest of the new world recently discovered. John Berger, when analysing Hans Holbein's *Ambassadors* (1533), observed that:

> To colonize a land it was necessary to convert its people to Christianity and accounting, and thus to prove to them that European civilization was the most advanced in the world. Its art included.[47]

Towards the end of the sixteenth century the disengagement of knowledge from religion was becoming more evident, and the advances in astronomy, physics and mathematics were changing the way space and time were perceived and conceived, paving the way to the development of the modern ocular-centrism of the Enlightenment. No one personified this new 'age of reason' better than the French philosopher and scientist René Descartes, a proponent of rationalism based on his methodological

[45] Harvey, David (1989). *The Condition of Postmodernity*. Oxford: Basil Blackwell, p. 245.
[46] Jay, Martin (1994). Op. cit., p. 57.
[47] Berger, John (1973). Op. cit., pp. 75–76.

doubt on the search for truth. His name is associated with the Cartesian coordinate system, the Cartesian geometry (analytical geometry) and Cartesian dualism, the separation of the body and the mind, the matter and the consciousness. His pursuit of logical reasoning was also directed to elaborate theories of vision and optics, which made him to be identified as a 'visual philosopher' and the 'founding father of the modern visualist paradigm'.[48]

In his essay *La Dioptrique* or *Optics*, Descartes affirmed that sight is the noblest of the human senses and that it is the mind, not the eye, that really 'sees', as images are formed in the brain like a mirrored reality of the world observed.[49] Vision, for Descartes, was the primary way to spatial perception, and he related the mechanism of creating images in the brain with the technique of perspective used in painting: in the same way the painter distorts shapes in order to create the illusion of three-dimensional space (for instance, representing circles by ovals), the brain reinterprets what the eyes capture of the external reality. Such Cartesian thought on sight and perception would be criticised in the twentieth century by phenomenologists such as Merleau-Ponty, who challenged Descartes' disembodied eye:

> Through phenomenological reflection I discover vision, not as a 'thinking about seeing', to use Descartes' expression, but as a gaze at grips with a visible world, and that is why for me there can be another's gaze; that expressive instrument called a face can carry an existence, as my own existence is carried by my body, that knowledge-acquiring apparatus.[50]

The re-engagement with the body underpins Merleau-Ponty's theory of perception; 'our own body is in the world as the heart is in the organism' he writes,[51] implying that one senses the world being *in* it, rather than looking *upon* it, as an analytical observer of the 'spectacle of the world', to use Merleau-Ponty's expression. As discussed earlier, Certeau strongly resonates this idea in his observations about the superior gaze above the city.

Cartesian dualism and its detached eye would, however, endure a long period of dominance in Western philosophy and arts, characterised by the

[48] Jay, Martin (1994). Op. cit., pp. 69–70.
[49] Jay, Martin (1994). Op. cit., pp. 75.
[50] Merleau-Ponty, Maurice (2002). Op. cit., p. 409.
[51] Merleau-Ponty, Maurice (2002). Op. cit., p. 235.

'scopic regime'.[52] Royal absolutism in France in the seventeenth century was intrinsically associated with visual representation that by all means enhanced the superiority of Louis XIV and his court. One striking example of visual documentation of this period is a grandiose map of Paris, elaborated in 1652 by the cartographer Jacques Gomboust. This map is considered to be the first detailed geometric survey of the city, produced according to the most up-to-date techniques, and as examined by philosopher and art critic Louis Marin, it was intended to depict the capital city as viewed from the position of the 'king's eye', who beholds the land, its buildings and its people, in a statement of a totalitarian power, often regarded 'divine'.[53] Decades later, in 1682, another carefully surveyed map was created for London by William Morgan, which presents a city that, despite being hit by a terrible disaster (the Great Fire of 1666), emerges as an ordered and magnificent place. Morgan's map, as its Parisian contemporary, makes use of high-end cartographic technology, and at the same time it celebrates the symbolic and political power of the urban landscape, which in the case of London is achieved by including a superb panorama view at the bottom of the map, along with images of the king and the great buildings of the city.

Towards the end of the eighteenth century in Europe, the romantic landscape emerges, assimilating the infinity of space by depicting a limitless horizon, 'where it seems possible to have a glimpse of the other side of the sky, a *beyond-space*'.[54] This undefined and blurred infinite space brings into the open possible 'places-to-be', as spatial utopias:

> It should be noted that the term *horizon* becomes infinite at the very moment when eighteenth-century dictionaries admitted *utopia* as a common noun. The terms *horizon* (like an infinite frontier?) and *utopia* (as the generic name of a textual, visual, literary, political generality) are related to each other through the historical and semantic relationship between the infinite and universality, and because of such a relationship are separated by a strange frontier.[55]

[52] Jay, Martin (1994). Op. cit., pp. 149–150.

[53] Marin, Louis (1988). *Portrait of the King*, trans. Martha M. Houle. Minneapolis: University of Minnesota Press.

[54] Marin, Louis (1993). Frontier of Utopia: past and present. *Critical Inquiry*, vol. 19, n. 3 (spring) pp. 397–420, p. 407.

[55] Marin, Louis. (1993). Op. cit., p. 408.

This relationship suggests ambiguity: to be able to gaze at the horizon, one needs to be placed somewhere, to establish a fixed point of view. Borders, or frontiers, must be identified by the observer for the purpose of creating distinctions, enabling us to relate 'here' and 'there', or 'now' and 'then'. Thomas More coined the term 'utopia', which gives the title of his book published in 1516, combining the Greek suffix 'u' (not) to the word 'topia' (place), which translates as 'no-place'.[56] His invented island is nowhere in the present but remains a possibility in a yet-to-come future—it presents a temporal horizon. As an ideology utopia is also associated with totalising views of the world, and uncoincidentally the map is the privileged means of representation:

> Utopia as ideology is a totality; and when political power seizes it, it becomes a totalitarian whole. The utopian representation always takes the figure, the form, of a map.[57]

These different ways of seeing—from the perspectivism of the Renaissance to the visual paradigm of the external observer that perceives the world from a distance, that was dominant in the seventeenth and eighteenth centuries—would have profound implications in the modes of representing the world in the European cartographic tradition, as will be discussed in the next section. The crisis of the 'scopic regime' in the nineteenth century, with the emergence of models of 'subjective grounded in the 'materiality of the body', as art critic Jonathan Crary puts it,[58] would trigger dramatic changes in the sciences and mostly in visual arts, with the advent of Impressionism and consequent art movements that dismantled the foundations of ocular-centrism. This would similarly have consequences for cartography: firstly with the creation of a modern mapping art and, secondly with the emergence of a critical school of geography that embraced the philosophical thought of phenomenologists and postmodernists.

[56] As an English word 'utopia' is a neologism but it has some correspondence with the Greek word *eutopia* ('good place') and *outopia* ('no place'). Marin observes that three names for More's Island surrounds the text: 'Outopia', 'Eutopia' and 'Oudepotia'. Marin, Louis (1993). Op. cit., p. 417.

[57] Marin, Louis (1993). Op. cit., p. 413; see also Marin, Louis (1984) *Utopics: spatial play*. New Jersey: Macmillan.

[58] Crary, Jonathan. Attention and Modernity in the Nineteenth Century. Jones, Caroline A.; Galison, Peter (eds.) (1998). *Picturing Science, Producing Art*. New York: Routlegde, p. 475.

Cartography, Knowledge and Reality

One of the main themes in the history of cartography is to define what is a 'map' and to identify the meanings associated with the concept of 'map' in different cultures and periods. The word *map* has a correlation with *plan* or *surface*. Throughout history maps have been designed, inscribed and marked in a variety of materials—stone, wood, paper, metal, fabric and film. In European modern languages the words *map* and *chart* are derived from the Latin *mappa* and *carta*, respectively. *Carta* denotes a formal document usually on a parchment or paper, while *mappa* indicates using cloth. It is interesting to note that *chart* in modern English refers more to maps of seas (used in navigation), while *map* is associated with the land and the skies (used in aviation). As Harley and Woodward recognised, in the preface of the first volume of their lengthy and ambitious project *The History of Cartography*, 'in a sense the subject has become a prisoner of its own etymology', and ambiguities carried in words such as 'map' and 'chart' persist to this day.[59]

The challenge posed by the definition of 'map' persists the distinctions between 'mapmaker' and 'cartographer', raising questions about how to categorise the immense variety of artefacts and practices that are linked in one way or another with registering a landscape. The word 'cartography'[60] is a neologism from the mid-nineteenth century created to refer to the study of early maps but since then has been used to include mapmaking, a practice often described as sitting between art and science.[61] Cartography involves science and survey techniques as much as aesthetical decisions to make our physical environment legible, along with communicating spatial information through graphic visualisations—or 'maps'.

[59] Harley, J. B. and Woodward, David (eds.) (1987). *The history of cartography: cartography in Prehistoric, Ancient, and Medieval Europe and the Mediterranean.* (HOC, Volume 1). Chicago: The University of Chicago Press. (Preface, pp. xvi–xvii). See also Harvey, Paul D. A. (1980). *The history of topographical maps: symbols, pictures and* surveys. London: Thames and Hudson. (p. 10).

[60] From Greek *khártēs*, meaning paper, chart or map, and *graphos*, meaning writing, describing. In English 'most of the current words in*-graph* are of the technical order and usually denote a thing that records or expresses (as if in writing) [...] Some words in*-graphy* denote processes or styles of writing or graphic representation, as *calligraphy*, *orthography*, *photography*, *typography*; but mostly they are names of sciences, as *bibliography*, *geography*, *lexicography*, *topography*'. Hoad, T. (1996). *The Concise Oxford Dictionary of English Etymology*.

[61] Harley, J. B. and Woodward, David (eds.) (1987). Op. cit., p. xvii.

The understanding of the map as an objective device for representing reality, disconnected from historical and cultural frameworks, and of mapping being a scientific and nonpartisan practice, has been confronted in the past decades by academics and artists alike. In scholarly circles the work of the geographer John Brian Harley from the late 1980s onwards is widely cited as a watershed on the critique of the power of maps and traditional mapping.[62] In his now classic essay *Deconstructing the map*, Harley questions why maps throughout history became a consensus of 'scientific' or 'objective' knowledge and how historians should challenge the 'epistemological myth' of cumulative scientific progress of representing the world ever more accurately and error-free.[63] This text appears just before a period when geographic information systems (GIS) were becoming more popular and accessible to the general public while also attracting much criticism.

Knowledge in cartography can be understood in an ample spectrum of epistemologies; on one end it is firmly based on the belief that the world can be surveyed and represented as it is and the validity of this knowledge depends solely on the progressive development of techniques and technology. At the other end, this knowledge is conditional: what we know from surveying our world is relative, subjective and contextual, and a universal 'truth' cannot be reached, as the reality is multiple and can be only understood within its multiple contexts. These contrasting approaches are characteristic of the recent history of cartography in the twentieth century and beyond, a history imbued with technological advancements seeking accuracy and veracity, as well as to contentious and disruptive mapmaking practices that refuse to confine mapping to conventional paradigms of universal 'truth', scientific precision and irrefutable representation of reality.

Amidst this debate, the concept of 'accuracy' is probably the most controversial subject in cartography and most problematic. Basically, accuracy involves handling and measuring degrees of certainty to which a map is a faithful representation of reality; the more 'accurate' a map is, the more 'correct' and closer it is to reality. However, this definition unravels a

[62] References and commentary of Harley's work can be found in Wood, Denis (1992). *The power of maps*. New York: The Guildford Press; Cosgrove, Denis (1999). *Mappings*. London: Reaktion Books; Pickles, John (2004). *A history of spaces: cartographic reason, mapping and the geo-coded world*. London, New York: Routledge; Crampton, Jeremy (2010). *Mapping: a critical introduction to cartography and GIS*. Oxford: Wiley-Blackwell.

[63] Harley, John B. (1989). Deconstructing the map. *Cartographica*, v. 26, n. 2, pp. 1–20.

string of questions: What is reality? What are the standards to measure certainty? How can representations be accounted as faithful or not? As Michael Goodchild points out:

> This notion of the map as a representation, or as a compilation of measurements, embodies a scientific perspective on cartography that is at odds with many alternatives, and implies that all aspects of the mapmaking process can be defined to rigorous standards of objectivity.[64]

Questioning the 'truth' of maps defies the belief that there is only *one* version of reality available to be captured and represented. One theoretical approach to deal with this problem is to consider that knowledge never comes from one single viewpoint; instead, it is situated—geographically, temporally, socially and ideologically. Therefore, there is no value-free or context-free knowledge.

Cartography in the 1960s and 1970s did not escape the influence of communication models for information transmission. Under the premise that information was basically transmitted from a sender to a receiver through a medium, communicating spatial information could be understood as a matter of building cartographic codes that could be shared between the cartographer (sender) and map reader (receiver). As far as this code, or language, was common to all involved, the 'transmission' would be successful; the problem is that such a code is far from being universal and most of all immutable, as the sender and the receiver do not process information in a constant, accurate manner. Certainly, this functionalist understanding of cartography was being fuelled by the possibilities that the then emerging field of GIS offered, but it was not difficult to realise that more diffuse, less precise generalisations about reality would continue to play a role in spatial representations, in maps.

Maps can be used as tools to make sense of data while relying on the overall perception of reality, and like diagrams and schemes they serve as compendia of information that summarises (although simplifying) the world around us. Looking at a map is an exercise of imagination as much as of interpretation, and to be fully understood, it is necessary to know their rules of representation, their own 'alphabet' and 'grammar' that

[64] Goodchild, Michael F. (2015). Accuracy. In Monmonier, M. *The history of cartography: cartography in the twentieth century* (Vol. VI). Chicago, London: The University of Chicago Press, p. 13.

compose spatial meaning. When used as literal instruments of information, maps are structured through the lens of mathematical algorithms that render a 'geocoded world', as the British geographer John Pickles emphasises.[65]

Representation and the Crisis of Representation

Representation has been a central theme in the advent of modern Western philosophy, explored by Descartes, Leibniz and Kant, among others. As presented in the previous section, the Cartesian model in particular influenced paradigms of visuality, considering vision a central part of scientific knowledge. The concept of representation as means to understand the world is fundamental to the modernist project. Edward Casey points out that 'representation' is a problematic term 'not in the fact *that* it happens [...] but in *how* it happens'.[66] Therefore, the question relies on how representations can be reliable and truthful when considered in the context of historical and political forces. Maps *re-present* these forces in a very particular way, as proxies standing on the place of the territory, constructing images of spaces that *re-implace* reality, to use Casey's expression.[67]

The 'crisis of representation' in human sciences emerges in the early 1980s contesting the dogmas of modernism, as signs of a broad reassessment of dominant ideas started to appear across disciplines, from law to architecture and the natural sciences.[68] John Pickles reasserts that 'profound epistemological changes' challenged the orthodoxies of sciences and humanities, resulting in 'nothing short of a *crisis of representation*'.[69] Factual descriptions and mimetic depictions of the world are put in doubt—accurate or not they are projections of human values and attitudes

[65] Pickles, John (2003). A history of spaces: cartographic reason, mapping and the geocoded world. London: Routledge.

[66] Casey, Edward (2002). *Representing Places: landscape paintings and maps*. Minneapolis: University of Minnesota Press, p. xvi (emphasis in the original).

[67] Casey refers to the 'phenomenon of re-implacement' as the ways in which 'places are altered and transmuted even as they are reinstated in painting and maps'. In this sense, maps are understood as representations that re-implaces the observed world every time they 'transmute' a place into cartographic form. Casey, Edward (2002). Op. cit., p. xv.

[68] Marcus, George E. and Fischer, Michael M. (1986) *Anthropology as cultural critique: an experimental moment in the human sciences*. Chicago and London: University of Chicago Press, pp. 7–8.

[69] Pickles, John (2003). Op. cit., p. 27.

towards material reality. Re-presentations are re-enactments of the world, and these will change over time: imagery is above all a cultural product, reflecting historical circumstances and ideologies.

The foundations of the 'traditional' modes of representation, and of how the world is imagined and represented in maps, are settled in the early modern period. During the Renaissance and the European age of discovery, maps acquired a key political and cultural role in documenting the exploration and colonisation of new worlds. Cartographic art was then employed to illustrate the discoveries as the ultimate human enterprise, as much as cartographic science would offer the tools for navigation. Map projection, technical perspective and new communication media come together in the fifteenth century in Western Europe, starting a long period of colonialism and the beginning of globalisation. Denis Wood considers that this is the period when modern maps or 'maps as we know them'[70] were born, at the service of the emergent European modern nation states that were applying the scientific revolution in astronomy initiated by Copernicus to their economic advantage.

The fifteenth century also witnessed the emergence of artists with a distinctive scientific interest, the most well-known being Leon Battista Alberti (1404–1472) and Leonardo da Vinci (1452–1519). Both are considered archetypes of the Renaissance polymath, venturing into architecture, painting, sculpture, engineering and literature among other professions, and both worked and lived in Italy, the cradle of the European Renaissance. Alberti founded his theories in mathematical concepts and provided the theoretical foundations of modern perspective geometry. Alberti's *reticolato*, also known as Alberti's 'grid' or 'grill' (*graticola*), is a mechanical aid to painters in the execution of the *finestra aperta* ('open window') technique, which involved creating a grid across an actual window in order to enable the artist to transfer the scene visible through the window to a correspondingly gridded canvas. In *Della Pittura*, considered the first treatise on painting, Alberti considered that the artist must produce a picture that will represent the visible world as if the observer of the picture were looking through a window (this being the frame of the final painting):

> [...] the picture must be so painted that a spectator's imagination is drawn towards the wall-plane, not away from it. This is why Renaissance painters,

[70] Wood, Denis (2010). *Rethinking the power of maps.* New York: The Guilford Press, p. 8.

> acting as interior decorators, revived and elaborated the system of perspective already used by interior decorators at Pompeii and elsewhere in the ancient world.[71]

Leonardo da Vinci's cartographic masterpiece is the *Plan of Imola* (1502), considered a turning point in city views, being the first geometric plan of a city and breaking with a tradition of symbolic maps inflated with religious and mystical features. In this map Da Vinci made use of the most advanced mapping techniques of his time, while adding an innovative approach, as the historian Paul Harvey observes:

> In many ways Da Vinci anticipated the developments in topographical mapping that occurred throughout Europe in the sixteenth century: in his concern for scale, his move away from pictorial representation towards outline plan, the union in himself of the surveyor and military engineer with the artist.[72]

When Da Vinci was commissioned by Cesare Borgia in 1502 to produce a detailed map of Imola, he had already developed the odometer and magnetic compass, which probably were put in use to measure and position all the buildings, streets and landmarks in the city. Da Vinci surveyed and drew a precise and accurate plan as seen from above, without the artistic flourishes and imprecisions of the classical bird's-eye views that were predominant at that time. Despite its innovative techniques and an original visualisation, the Imola map might have benefited from previous surveys and older maps. Successive transpositions of information and design, in a 'copy-and-paste' fashion, are a common feature in the history of cartography. Nevertheless, it marks a transition from mappings of the fantastic and mythical to a mapping of information and science.

The quest for the perfect, realistic representation of the world continued throughout the sixteenth century. In 1569 the Belgian cartographer and geographer Gerardus Mercator presented a map of the whole globe that was recognised as giving the most accurate and correct representation of the Earth's surface until then—in fact, Mercator's projection became a

[71] Collingwood, R. G. (1938) *The Principles of Art.* Oxford: Oxford University Press.

[72] Harvey, Paul D.A. (1980). *The history of topographical maps: symbols, pictures and* surveys. London: Thames and Hudson, p. 155.

standard for maps, and it is still used to this day.[73] According to Ballon and Friedman there was an 'explosion in the production of images of cities' in the Renaissance as they became the central sites of political, cultural and economic power in Western Europe, being also used as instruments of propaganda.[74] Already in the early fifteenth century, mural frescoes of townscapes decorated palaces, and later, with the invention of the printing press by Guttenberg about 1439, city maps illustrated *incunabula* produced in Germany and Italy.[75] Map printing, using relief and intaglio techniques, were well developed in China since the twelfth century, but the volume and character of this production cannot be compared with that of Western Europe in the Renaissance. In China printed maps were produced in small numbers for administrative and educational purposes, while in Europe printed maps became a commercial endeavour, produced in large numbers to attend the demands of the state, scholars and merchants alike.[76]

The early industry of printed maps was based on reproductions made by means of woodcut or engraving techniques, resulting in expensive and large-scale loose sheets that were not widely available. However, in 1570 a cartographer from Antwerp, Abraham Ortelius, launched a new product in the market: the *Theatrum Orbis Terrarum* (Theatre of the world), a compendium of maps in book form, which was more affordable and easier to handle. Now considered the first modern atlas, it brought together maps arranged by continents and regions, with descriptions and references to cartographers and places, in a legible and organised structure. This formula was repeated by editor Georg Braun and engraver Franz Hogenberg two years later in Cologne with the publication of *Civitates orbis terrarum* (Cities of the world, 1572), an extensive collection of views of towns from all over the world. The *Civitates* became an instant success and other

[73] Google Maps uses a variant called 'Google Web Mercator'. The projection method, while distorting the representation of global landmasses, is still widely applied in the small scale (up to 1:50.000) in local maps.

[74] Ballon, H. and Friedman, D. Portraying the city in early modern Europe: measurement, representation and planning. Woodward, David (ed.) (2007). *The history of cartography: cartography in the European Renaissance.* Volume 3. Chicago, University of Chicago Press, pp. 680–704.

[75] Iuliano, Marco. Melchior Lorck's Istanbul in the European context. Fischer, Erik (ed.) (2009). *Melchior Lorck.* Copenhagen: Vandkunsten Publishers, vol. 4, pp. 25–60; p. 25.

[76] Woodward, David. Techniques of Map Engraving, Printing, and Colouring in the European Renaissance. Woodward, David (ed.) (2007). Op. cit., p. 591.

volumes followed; the sixth and last of the series was published in 1617.[77] Georg Braun claimed for himself the creation of the original idea, but other attempts had been made previously. Nevertheless, he is recognised as establishing a new standard for city atlases, skilfully combining imagery and commentary, as scholar Lucia Nuti observes:

> The commentary fully mirrors the double nature of the work, suspended in a perfect balance between market and science [...] He goes through the major scholarly themes, as they had emerged from more than a century of debate: the primacy of the city in human history, the primacy of architecture among the arts, and particularly the primacy of sight among the senses.[78]

Braun and Hogenberg gathered images that conveyed stylistic uniformity to their atlas, giving preference to bird's-eye views of the cities depicted. The arrangement of the plates, the size of the maps and the decorative elements such as margins and the coat of arms were all carefully combined to reinforce the concepts implicit in the descriptions: civility, order and entrepreneurship. *Civitates orbis terrarum* is a celebration of urbanity according to the modern sixteenth-century European paradigm.

The rediscovery of Ptolemy's *Geographia*, translated into Latin in the early fifteenth century and later published in Florence, had a strong influence on the way cities were represented in maps, including the work of Braun and Hogenberg. Alongside Brunelleschi's revival of perspective, Ptolemy's treatise gave the foundations and inspiration for a more topographical style to urban views, first developed in Italy and then spreading over Europe.[79] Ptolemy also distinguished two types of cartography: the *chorography*, or the description of places, and *geography* as the description of the entire world and the mathematical relations between its different parts.

Mapmaking and perspective in the Renaissance emerged as artistic devices to describe landscape. Maniglier notes that 'both manifest a historical shift from a *symbolic* to a *representational* conception of space', meaning that the image is not used to express a thought but rather represent another space. However, this representation is not mimetic, as

[77] *Civitates orbis terrarum* was the tile of the first volume only but the whole collection is commonly known by it.

[78] Nuti, Lucia. (1994). The perspective plan in the sixteenth century; the invention of a representational language. *The Art Bulletin*, 76, 1, pp. 105–128; p. 106.

[79] Iuliano, Marco (2009). Op. cit., pp. 28–30.

mapmaking and perspective 'substitute the notion of *resemblance* with that of *projection* of one space into another one'.[80]

Against the claim of an unequivocal hegemony of the Italian perspectivism, Svetlana Alpers argues that the Dutch art of the seventeenth century established another visual paradigm in classical painting that was closer to cartographic representation.[81] According to Alpers, Dutch painters conceived a more objective and neutral form of mapping that informs about the world, while the Italian perspective was motivated by telling (or retelling) stories. Alpers also makes a distinction between maps and landscapes:

> [...] maps give us the measure of a place and the relationship between places, quantifiable data, while landscape pictures are evocative, and aim rather to give us the quality of a place of the viewer's sense of it. One is closer to science, the other to art.[82]

Topographical[83] (or topographic) maps basically depict the shapes and features of land surfaces, are usually large scale and differ, for instance, from the often small-scale maps showing political and administrative regions. On trying to represent land surfaces topographical maps resemble aerial views while bearing information that is not visible at all in the actual landscape, such as place names and contour lines. Besides, they reproduce features in a simplified graphic version: a house can be represented as a single square, used as a symbol. These characteristics are intrinsic to maps and not to photographs, also because a map brings a vertical and uniform representation of the land at any point, while a photograph will show the typical distortion in shape and scale away from the point it was taken.

The topographical map was not an invention of the sixteenth century, but it was then that the combination of mapping and surveying turned into a tradition that dominated cartography, setting up a mode of representation that survives until our days. According to Paul Harvey, in the Middle Ages surveyors and map makers worked separately: the first being

[80] Maniglier, Patrice. Maps in perspective; what can philosophy learn from experimental maps in contemporary art? Levy, Jacques (2015). *Cartographic Turn*. Oxford, Routledge, p. 46.

[81] Alpers, Svetlana (1983). *The art of describing: Dutch art in the seventeenth century*. Chicago: University of Chicago Press.

[82] Alpers, Svetlana (1983). Op. cit., p. 124.

[83] Related to 'topography', meaning the study and description of the features of a locality; the word derives from the Greek *topos* (place) and *graphia* (writing).

responsible for measuring the land (and not necessarily drawing a map of it) and the second for creating pictorial representations of the land, which Harvey calls picture-maps.[84] Usually, map makers were artists working in pictures of landscape—they were primarily painters and engravers—while surveyors had their training based on military engineering.

The best and most prolific group of mapmaking artists were concentrated in Renaissance Europe, particularly in Italy. Detailed plans and bird's-eye views of cities are known to have been made since the twelfth century, but it was only by the late fifteenth century they became much more realistic, initiating a new tradition of picture-maps. Two exemplary works of this development are a view of Florence engraved by Francesco Rosselli (circa 1485) and of Venice by Jacopo de Barbari (1500). Both Italian artists excelled in producing vivid representations of urban landscapes using direct observation, perspective system and measuring techniques to compose intricate bird's-eye views.[85]

These town views, like the ones that appear in the later editions of Braun's and Hogenberg's *Civitates*, are the product of a union between two representational modes, the geometrical and the pictorial, in a trend to create images as close as possible to the observable reality. Through the sixteenth century this led to the appearance of the topographical map drawn to scale and, as Harley comments, to pass the work of mapmaking from the hands of the artist to the hands of the surveyor by the end of the sixteenth century.[86]

Printed town views and town plans appeared relatively late in England, according to cartography historian Sarah Bendall.[87] In Britain a similar project to the *Civitates*, an atlas of regions and cities produced by John Speed, the *Theatre of the Empire of Great Britaine*, was only published in 1611, almost 40 years after the German book. New instruments and techniques for surveying the land continued to be developed throughout the seventeenth century and paved the way to the foundation of modern cartography in the eighteenth century. The great improvement in surveying methods facilitated the emergence of the systematic topographic mapping of whole countries, as an enterprise conducted by nation states. The UK

[84] Harvey, Paul D.A. (1980). Op. cit., p. 14.

[85] Nuti, Lucia (1994). Op. cit., pp. 115–116.

[86] Harley, Paul D. A. (1980). Op. cit., p. 164.

[87] Bendall, Sarah (2002) Draft towns maps for John Speed's 'Theatre of the Empire of Great Britaine'. *Imago Mundi*, vol. 54, pp. 30–45.

and France were pioneers, with the creation of their national survey systems.[88] The rationalist culture of the Enlightenment removed all traces of pictorial language in maps and adopted the exactitude of lifelikeness representation.[89] The city map moved from portraiture to geometric precision, becoming a functional tool and ever more oriented to support local government. Town plans were produced seeking a high degree of accuracy to meet the demands of public bodies occupied with planning and administration.

Meanwhile more ludic representations of cities experienced a revival in the last decades of the eighteenth century in Europe in the form of panoramas. The panorama has an interesting history as an invention with a patent, registered in 1787 by Irish painter Robert Barker, describing the principles and features of his 'coup d'oeil' technique and how to properly display paintings in a 360-degree canvas. Barker first exhibited his creations in London in 1792, when he coined the definite term 'panorama', in a rotunda specially built to enhance the illusionary effect of the paintings. The panorama as entertainment and spectacle became very popular throughout the eighteenth century; companies were formed to specifically produce canvases for great exhibitions and the world's fairs.[90] Barker had not only invented a new painting style but also a new way of viewing, and at its conception, the panorama had brought a new way of viewing cities. Walter Benjamin reflected on panoramas and other inventions that followed in the nineteenth century, like dioramas and ***kaisepanoramas***, as spaces where illusion and reality converge:

> The interest of the panorama is in seeing the true city – the city indoors. What stands within the windowless house is the true [*sic*]. Moreover, the arcade, too, is a windowless house [...] What is true has no windows; nowhere does the true look out to the universe.[91]

The pursuit of the ideal representation by the mapmakers of the sixteenth and seventeenth centuries transformed the history of cartography through their attempt to apprehend the world as a whole in graphic form. The Renaissance maps appeal to the eye while applying techniques of

[88] The Ordnance Survey and the Institut Géographique National, respectively.

[89] Nuti, Lucia. (1994). Op. cit., p. 120.

[90] Comment, Bernard (1999). *The Panorama*. London: Reaktion.

[91] Benjamin, Walter, (2002). *The Arcades Project*, ed. R. Tiedemann, Cambridge, MA: The Belknap Press of Harvard University Press, p. 532.

measurement and survey first theorised by Ptolemy in his *Geographia*. This attempt to get closer to observable reality and reach aesthetic pleasure results in deceitful images that mimic landscapes that cannot be seen in the real world. Travellers of the sixteenth and seventeenth centuries visiting Italian cities searched in vain to experience the same views depicted in their books and atlases, only to find out that such views were in fact clever illusions.[92] They are patchworks of individual drawings, making use of the manipulation of perspective and scale. Such artistic techniques to produce bird's-eye views were never abandoned, and they continue to be used to this day, also in digital cartography.

Contesting Traditional Cartographies

The modes of cartographic representation discussed in the previous section showed that in the establishment of official, state-led or corporate-led, mapping practices were strongly based on epistemologies that reiterate a direct correspondence between representation and truth. As far as the 'crisis of representation' is concerned with cartography, these practices became increasingly related to 'traditional cartography' towards the end of the twentieth century. Most of the criticism they received pointed to discourses of power and dominance, technical and political, that supressed alternative representations. Positivism and quantitative methods that dominated this approach in cartography were targeted as tending to consider maps as ways to present a 'fixed' and 'planar' world, deprived of its social-temporal dynamics, as if in an inert landscape.

In the early 1970s, after a decade emblematised by the Cold War, counterculture, political protest and space travel and exploration, contestation and criticism over the totalising, disembodied and the male gaze upon the world came forth through the work of Anglo-American cultural geographers and social scientists. Belonging to a generation who were born around or just after the Second World War, these scholars engaged with the study of space and spatiality embracing postmodernist, postcolonial and feminist theories, producing an influential body of work throughout

[92] Nuti, Lucia. (1994). The perspective plan in the sixteenth century; the invention of a representational language. *The Art Bulletin*, 76, 1, pp. 105–128; p. 128.

the last decades of the twentieth century that became known under the label of 'critical geography' or 'radical geography'.[93]

Two of the fiercest critics of scientific mapmaking and the cultural biases infused in traditional cartography are the British geographer John Brian Harley and American scholar Denis Wood. Harley produced a series of polemical papers in the mid-1980s, culminating with his most influential essay *Deconstructing the map*, published in 1989, and Wood published his seminal book *The Power of Maps* in 1992. They are both inspired by French theorists Michel Foucault and Jacques Derrida, from whom they drew debates on power and knowledge relationships (Wood, after Foucault) and deconstructivist approaches to mapping (Harley, after Derrida and Foucault). This line of thought resulted in a critical work that acknowledges the potentialities for transformation that mapmaking can induce while reinforcing the understanding that maps are ideological constructs and not neutral depictions of reality. The new theories attacking old epistemologies are characterised as 'post-paradigms', such as postmodernism, poststructuralism and post-Marxism, and critical cartography, inserted into this context of 'post-isms', attacked the paradigms of scientific representation of precision and accuracy, stressing that certainty is not achievable or consensual. As Denis Wood advocates:

> Far from being pictures of the world maps are instruments for its creation, they are not representations but systems of propositions, arguments about what the world might be.[94]

Most of the ideas surrounding the critique of cartography were placed in opposition to, or against, what was considered 'traditional', 'cartesian' and 'immutable' cartography. As Del Casino and Hanna (2006) pointed, scholars writing in the 1980s–1990s, including geographers J. B. Harley and Denis Wood, maintained 'an implicit duality' between maps as representations of space and spatial practices, focusing either on the social, political and economic contexts of production or on the consumption and use of maps. These approaches create a separation between author and

[93] Peet, Richard (2000). Celebrating thirty years of radical geography. *Environment and Planning A*, 32, 6, pp. 951–953.

[94] Wood, Denis (2010). *Rethinking the power of maps*. New York: The Guilford Press, p. 8.

reader, object and subject, design and use,[95] and ultimately disconnect maps from spaces. Del Casino and Hanna argued that maps and mappings are simultaneously representations and practices.

Other scholars such as David Harvey criticised cartography because it 'imposes spatial order on phenomena' and 'in its contemporary manifestation ...depends heavily upon a Cartesian logic in which *res extensa* are presumed to be quite separate from the realms of mind and thought and capable of full depiction within some set of coordinates'. For Harvey 'how urban life is experienced and practiced [...] has much to do with how we form and reform mental maps of the city'.[96] His fellow British geographer Doreen Massey has equally challenged conventional notions of cartography as a purely spatial category. Maps are not simply representations of space but of space and time, she affirms, and cartographic representations are thus connected to 'space as a sphere of dynamic simultaneity, constantly disconnected by new arrivals, constantly waiting to be determined (and therefore always undetermined) by the construction of new relations'.[97]

The wave of criticism about positivist views in geography was also aimed at the tools and methods associated with them, especially those related to geographical information systems. Although GIS has been around since the 1960s, it was during the 1990s that an increasing 'friction' between its practitioners and human geographers was observed.[98] A series of articles in academic journals ignited a debate around methodological and epistemological issues, with human geographers criticising the positivist approaches and Cartesian methods among GIS practitioners, and the latter pointing to the lack of understanding of GIS techniques of their counterparts. As examples, the comments below appeared in two articles published in 1989 in the same issue of the journal *Cartographica*; the first is from J. B. Harley and the second from Denis Wood:

[95] Del Casino, Vincent J. and Hanna, Stephen P. (2006). Beyond the 'binaries': a methodological intervention for interrogating maps as representational practices. *ACME: An International E-Journal for Critical Geographies,* 4 (1), 34–56.

[96] Harvey, David (2001). *Spaces of capital: towards a critical geography.* New York: Routledge, pp. 220–221.

[97] Massey, Doreen (2005). *For space.* London: Sage, p. 108.

[98] Schuurman, Nadine (2000). Trouble in the heartland: GIS and its critics in the 1990s. *Progress in Human Geography* 24, 4 pp. 569–590; p. 569.

> [...] as they embrace computer-assisted methods and Geographical Information Systems, the scientific rhetoric of map makers is becoming more strident. (Harley, 1989: 2)

> [...] computer graphics are in the service above all else of the military, and secondarily automobile manufacturing [...] that is, computer graphics mainly promote death, either directly, through the support of weapons system, or indirectly, through the use and production of cars. (Wood, 1989: 118–119)

In agreement with Harley and Wood was the American geographer Mark Monmonier who in his book *How to Lie with Maps* (1991) associates mapmaking with the means to exercise and enforce relations of power.[99] This initial debate about GIS was mostly centred in the critique of positivism and technocracy that led to a naïve position of practitioners, apparently disengaged from reality and unaware of the social consequences of applied technologies. Nevertheless, there was not much argument about 'whether or not GIS research [...] can necessarily be linked to its implementation'.[100] In fact, most of the critique was coming from outsiders, with little or no knowledge of practical GIS.

This scenario started to change from the mid-1990s, with critics and advocates of GIS joining efforts to debate the ethics and the social nature of utilising geospatial technologies. In an article published in 1995, Eric Shepard summarises:

> [...] the fundamental question is whether the logic of GIS, as a result of design decisions [...] privileges certain views of the world over others [...] even if GIS were available to everyone who owned a computer, it would still not be a democratized technology as it would continue to embody algorithmic thinking, itself limited.[101]

Shepard is considering the dominance of private corporations and public agencies in the ownership of the technology and hardware capable to compute geographical data, but from 2000s GIS became largely available,

[99] Monmonier, Mark (1991) *How to lie with maps.* 2nd edn. Chicago: University of Chicago Press.

[100] Schuurman, Nadine (2000). Op. cit., p. 575.

[101] Shepard, Eric (1995). Sleeping with the enemy, or keeping the conversation going? *Environment and Planning A* 27, 1026–1028, p. 1027.

thanks to open-source software and personal computers that are more powerful. The possibilities for digital mapmaking and spatial data sharing increased exponentially since then, and cartography was no longer an exclusive domain of experts. This new outburst of cartographic practices is recognised as 'critical cartography', and again Anglo-American scholars dedicated themselves to study the phenomenon, such as British geographer Jeremy Crampton and American geographer John Krieger who, already in 2005, discussed how powerful business and governmental elites were being challenged by the newcomers, operating with collaborative tools and mobile applications.[102] The critical movement that emerged from this, they argue, was part of a long tradition of subverting the politics and representation of space, including artistic practices that involved 'map experimentation', with roots in avant-garde movements at the turn of the twentieth century.[103]

In a more phenomenological approach to mapping, British anthropologist Tim Ingold argues that to understand space one needs to move through it: 'we *know as we go*, from place to place'[104] he writes, implying that spatial knowledge is derived from a sensorial experience. The pedestrian here is again summoned to connect idealised spaces to experienced spaces. Ingold's proposition reflects in a critique to the 'modern cartography', or the 'scientific map' that in his view excludes the practices of exploration and observation:

> In the work of the modern cartographer knowledge generated through movement [...] is presented as if it issued from a totalising vision above and beyond the world [...] the imaginary 'bird's-eye view' of a transcendent consciousness.[105]

The duality viewing from above/viewing from the ground seems to persist in mapping practices. In a workshop organised in 2010 in Montreal, Canada, cartographers and artists alike were invited to develop a mapping project inspired by a database compiled by the students of the Masters of

[102] Crampton, Jeremy, and Krygier, John (2005). An Introduction to Critical Cartography. *ACME: An International Journal for Critical Geographies* 4 (1), 11–33.

[103] Crampton, J. and Krygier, J. (2005). Op. cit., p. 17.

[104] Ingold, Tim (2000). *The perception of the environment: Essays on livelihood, dwelling and skill*. London: Routledge. (p. 229) (in italic in the original).

[105] Ingold, Tim (2000). Op. cit., p. 230.

Environmental Impact Assessment (EIA) at Concordia University.[106] At the end of the event the organisers found 'a major dichotomy' in the way cartographers and artists approached mapping: while cartographers took the database as 'a source of information that can be mapped from above', the artists saw on it a source of 'inspiration to orient their research efforts from the ground':

> From an artistic perspective mapping is personal, subjective, associated to ground exploration and dedicated to conveying the complexity of the phenomena being studied. From a cartographic perspective, mapping is data dependant, as objective as possible, provides a view from above distanced form the phenomena mapped, and largely dedicated to simplify the complexity of the phenomena represented as much as possible.[107]

What is interesting to notice is the distinction between 'mapping from above' and 'mapping from the ground', as if the map maker's profile, being either a cartographer or an artist, is the true determinant to their worldview. One of the conclusions of the workshop held that the 'view from above' is a cartographic perspective and the 'view from the ground' is an artistic one,[108] reifying traditional divisions in human knowledge: practice and theory, technique and expression, science and art. However, as will be discussed in Chap. 4, contemporary art mapping practices are increasingly blurring such boundaries, appropriating tools from technical expertise and scholarly argumentation into aesthetic, social and political experimentation.

Performance, Agency and Deep Maps

During the second half of the twentieth century other fields of investigation evolved around concepts of space and spatial perception, and one of

[106] Caquard, Sebastian, Vaughan, Laurene and Cartwright, William (eds) (2011). *Mapping environmental issues in the city: Arts and Cartography cross-perspectives*. New York, Springer.

[107] Caquard et al (2011). Op. cit., p. 7.

[108] In the organisers' words: 'While in cartography, mapping refers to specific techniques and concepts for representing information related to places, in the arts and humanities this term is used more broadly and includes a diversity of practices and expectations. More than just data, artists and other creative practitioners express their understandings, fears, hopes, emotions and perceptions about places and people through mapping, and these dimensions are essential for fully understanding our relation to places'. Caquard et al, (2011). Op. cit., p. 1.

them that has particular repercussions in the study of cities was centred in the relationships between the physical space and mental maps. Building mental maps is a capacity that all humans have and animals alike. It goes down to instincts of self-preservation that humans learn to navigate the land, to find and remember the routes that lead to food and shelter or to avoid others where danger is present. Living in cities these more innate skills tend to be less evident and, of course, maps come in handy to support navigation. What happens then is that other skills are triggered and developed to negotiate the embodied knowledge acquired from exploring the environment and the codified knowledge acquired from maps. The enduring success of London's Tube map, first conceived in 1931 by Harry Beck, is an example of how the human brains process the cartographic language, finding a balance between accurate representation and simplification through geographical distortion.

Cognitive science emerged in the mid-1950s and entered the field of urban studies through the work of the architect Kevin Lynch in 1960, when he published his most well-known book *The Image of the City.*[109] Lynch's approach was close to structuralist methods, trying to establish what aspects of the urban environment were perceived and retained by the human brain and how people could 'read' the city from observation. Before Lynch the use of the term 'cognitive map' first appeared in 1948 in the paper *Cognitive maps in rats and humans*, from the psychologist Edward Tolman,[110] and later, in 1974, urban geographers Gould and White[111] developed further the notion of 'mental maps' applied to human spatial perception in relation to geographical location. According to Portugali (2018) these three concepts—Tolman's cognitive behaviour, Lynch's mental images and Gould and White's cognitive maps—established the foundations of cognitive geography.[112]

Lynch's study of the image of the city indicated that what is imprinted in our minds from observing and reacting to the urban landscape can be brought down to five basic elements: edges, nodes, paths, districts and landmarks. This system is similar to a code, or a legend in the map. In fact,

[109] Lynch, Kevin (1960). *The image of the city.* Cambridge, MA: MIT Press.

[110] Tolman, Eduard C. (1948). Cognitive maps in rats and men. *Psychology Review*, 55, pp. 189–208.

[111] Gould, Peter, and White, Rodney (1974). *Mental Maps.* Harmondsworth: Penguin.

[112] Portugali, Juval (2018). Cognition and the city: and introduction. *Built Environment*, vol. 44, n. 2, editorial.

Lynch's illustrations for this theory resemble diagrammatic drawings and emulate network graphs. Other researchers in the late 1970s explored the urban landscape and morphology in terms of semantic structure, such as Christopher Alexander's *pattern language* and *generative grammar*, at the University of California, Berkeley,[113] and Bill Hillier's *space syntax*,[114] at University College London.

These early theories in cognitive mapping are related as well as to the then emergent computing techniques of geographical information systems. The vocabulary of the urban designers and theorists is noticeably similar to those working in computing: nodes, networks, flows, categories, systems, models and patterns. In urban studies cognitive mapping was imbued with psychology's empirical methods, mathematics and computing science. 'The perception of a city is a transaction between person and place', wrote Kevin Lynch,[115] denoting the individual character of the human-environment relationship. Culture, society, economic situation and political system are all major factors that will influence one's perception, but ultimately it is down to psychological factors and particular circumstances in time and space that a perception is formed and often translated into some imagery, either visual or not. This ascertainment is not devoid of problems.

Reconsidering his own research 20 years later, Lynch clearly pointed out a few of them, along with criticisms that the study received at the time: the small sample size, using map drawing as the main method (not a familiar language for most of the users and researchers alike), appropriation of a creative skill (drawing) for emulating a 'science of design' and basic relevance.[116] However what Lynch highlights as a major problem in *The image of the city* is that it lacked an understanding of the variations in the individual observer in order to emphasise the variations in the physical space. That emphasis on spatial patterns and connectivity led planners and designers to adopt them as the new formula for urban design, moving

[113] Alexander, C., Ishikawa, S. and Silverstein, M. (1977). *A pattern language: towns, buildings, construction.* New York: Oxford University Press. See also: Alexander, C. (1979). *The timeless way of building.* New York: Oxford University Press.

[114] See Hillier, Bill and Hanson, Julienne (1984). *The social logic of space.* Cambridge: Cambridge University Press.

[115] Lynch, Kevin (1984). Reconsidering the image of the city. Rodwin, L. & Hollister, R.M. *Cities of the mind: images and themes of the city in the social* sciences. New York: Plenum Press, pp. 151–161.

[116] Lynch, Kevin (1984). Op. cit., p. 155.

away from the original aim of the study, which was to attract attention to the necessity of listening to the public, the city space users.

The maps that illustrate *The image of the city* have become an abridged version of the study and icons for an era in urban planning. Cosgrove and Martins (2000) pointed out that cognitive mapping, as studied in geographical and psychological research in the 1960s and 1970s, 'were not concerned to challenge directly formal cartography's capacity to create accurate spatial images'.[117] In fact, the mental maps proposed by Lynch were tools for urban design that enhanced the spontaneous experience of the observer—foremost, a pedestrian—in the city rather than take the arbitrary, 'from above' approach of traditional urban planning. It was a true attempt to humanise the science of the cities. Nevertheless, the more scientific elements of categorisation, hierarchy and functionality were incorporated in the maps to make them viable for planning and design.

In a well-known and often cited passage of *A thousand plateaus*, the joint oeuvre of Deleuze and Guattari, the philosophers use the opposition of mapping and tracing as one of the principles to develop their 'rhizomatic' theory.[118] Mapping, in their proposition, is analogue to a rhizome, while tracing reproduces the hierarchical, over-coded mode of thought:

> The rhizome is altogether different, *a map and not a tracing*. [...] What distinguishes the map from the tracing is that it is entirely oriented toward an experimentation in contact with the real. [...] A map has multiple entryways, as opposed to the tracing, which always comes back 'to the same'. The map has to do with *performance*, whereas the tracing always involves an alleged 'competence'.[119]

On considering the mapmaker's perceptions and observations, Deleuze and Guattari indicate an emphasis on the *practice* rather than in the *object*. Attention was shifted from maps to mappings, and in doing so, the cultural, political and social realms were instrumental to what should result as 'map'. This is what Crampton identifies as the performative character of mapping:

[117] Cosgrove, Denis and Martins, Luciana (2000). Millennial geographics. *Annals of the American Association of Geographers.* 90 (1) pp. 97–113; p. 107.

[118] Based on the botanical rhizome, Deleuze and Guattari theory advocates multiplicity and non-hierarchical thought.

[119] Deleuze, G. and Guattari, F. (1987) *A thousand plateaus: capitalism and schizophrenia.* Trans. by Brian Massumi. Minneapolis: University of Minnesota Press. pp. 12–13 (emphasis in the original).

> Performativity is a cultural, social and political activity; maps as protest and commentary [...] We seem to be moving from a niche-based study of maps as objects to a more comprehensive (and potentially interdisciplinary) study of mapping as practice, the knowledge it deploys, and the political field of its operations.[120]

Cosgrove and Martins also used the term 'performative mapping' back in 2000 on reflecting upon the millennial celebrations in Rome and London; for them the term refers to:

> [...] ways in which *genius loci*, the capacity of particular places to communicate intense and unique meaning across space and time, is actively made and remade [...] Performative mapping implies the contemporary convergence between artistic and scholarly activity in geography, simultaneously understanding and activating spatialities.[121]

This approach to mapmaking that differentiates itself from traditional cartography is mostly observed among artists who have being using, defacing, transforming and creating maps and ways of mapping. This more creative, skewed relationship between art and cartography, not circumscribed by representational techniques and useful purposes of orientation and navigation, goes back as far as the first decades of the twentieth century.[122]

As commented previously, towards the end of the 1990s, debates around GIS and technical mapping shift from fierce criticism to a more inclusive approach. A good example of this change is James Corner's essay *The agency of mapping*, published in 1999.[123] At the very beginning, Corner states that he is aware of the criticism and 'coercive' aspects of maps, being instruments of power and domination, but instead chooses to focus upon 'more optimistic revisions of mapping practices'.[124] His optimism is based on the potentiality of maps and mapping practices to 'uncover realities' rather than merely reproduce 'what is already known'.

[120] Crampton, Jeremy (2009). Cartography; performative, participatory, political. *Progress in Human Geography*, 33 (6), pp. 840–848; p. 840.

[121] Cosgrove, Denis and Martins, Luciana (2000). Op. cit., p. 97.

[122] Denis Wood (2006) and (2010) proceeded to an extensive catalogue of artists and art collectives over the period.

[123] Corner, James. The agency of mapping: speculation, critique and invention. In Cosgrove, Denis (1999). *Mappings*. London: Reaktion, pp. 213–300.

[124] Corner, James (1999). Op. cit., p. 213.

Being a landscape architect Corner relates the act of mapping to the act of drawing, proposition and design[125] while considering maps as both analogical and abstractive devices. His discussion centres on identifying four thematic ways in which designers and planners are using the creative capacities of mapmaking: 'drift', inspired by the situationists' *dérives*; 'layering', superimposing spatial elements and experiences; 'game-board', conceiving space as platform for contesting agents to act and play; and 'rhizome', based on the concept of Deleuze and Guattari of open-ended, indeterminate strategies.

Corner brings the situationists into the discussion as examples of the transformative power of mapping in unveiling connections and constructing meanings, not merely in support of gathering information or mimicking physical features. Following up the proposition of his four categories of mapmaking, Corner revisits plans and sketches of the postmodernist architecture of the 1980s–1990s, authored by the likes of Peter Eisenman, Raoul Bunschoten, Rem Koolhaas and Bernard Tschumi. The argument for the agency of mapping relies on making maps and using them within a conceptual framework that differs from the traditional analytical survey, which also has impacts in designing and planning cities. As Corner points out conventional maps are produced aiming to reflect the exactitude and accuracy that is expected from urban design in order to make decisions about the use and occupancy of the territory. However, design decisions are usually creative endeavours, and pre-design survey maps usually do not match this 'presumed inventiveness'.[126]

To consider the agency of mapping we should look again into the faculty of maps to translate the world into visual abstractions, to overcome analogue representations of reality to provide material that infers the realm of spatial imagination. This act of translation makes mapping also a generative tool for urban design; a recurrent example is that of *New Babylon*, the collaged maps created by the Dutch architect and situationist collaborator Constant Nieuwenhuys in the 1950s. Aarati Kanekar's analysis of drawings produced by architect Perry Kulper[127] echoes Corner's

[125] 'Design' and 'designate' share a common etymological root. Designate comes from Latin *designatus*, past participle of *designare* (in currently English 'design'), meaning 'mark out, devise, choose, appoint'. *Designare* is derived from *de*—'out' and *signum*—'identifying mark, sign'.

[126] Corner, James (1999). Op. cit., p. 216.

[127] Kanekar, Aarati (2015). *Architecture's pretexts: spaces of translation*. London and New York: Routledge, Taylor & Francis Group.

propositions for the use of maps in urban design. Writing 15 years later Kanekar (referencing Corner) considers those drawings as examples of metaphorical maps that are instrumental for design, evidencing the generative qualities—or agency—of mapping.

In Perry Kulper's proposal for Davids Island,[128] the juxtaposition of layers of textual and graphic information is combined with registers of physical features, events, military and nautical cartography and various notational systems. The techniques of map collage first experimented by the situationists are here employed to question the boundaries of space and time. Kulper depicts Davids Island past history, present features and future development combined in the same drawing, which Kulper describes as 'opening representational borders' and overcoming the legacy of traditional architectural drawings of 'reductive presentational practices'.[129] As Kanikar notes these are 'thick' drawings that 'augment patterns of relationships [...] an interesting amalgam of the existing visible, the inaccessible invisible and the projected imaginative'.[130]

Kulper's drawings reflect Corner's remarks about maps being 'in-between the virtual and the real', which in architecture practice can be related to drawings that are in between the visionary design and the concreteness of the actual territory. Corner elaborates on this concept of the exploratory nature of maps as a dynamic process:

> [...] [maps] permit a kind of excavation (downward) and extension (outward) to expose, reveal and construct latent possibilities. [...] Mapping is neither secondary nor representational but doubly operative: digging, finding and exposing on the one hand, and relating, connecting and structuring on the other.[131]

Thus, it is through the act of mapping, when one chooses the information to collect ('digging') and how to represent it ('connecting'), that new meanings, or 'new realities' as Corner puts it, are constructed and

[128] Davids Island, named after the ink manufacturer Thaddeus Davids, who purchased the land in the mid-nineteenth century, is a small island off the coast of New Rochelle, New York, USA.

[129] Kulper, Perry (2005). Representing beyond the surface. *arcCA 05.3—Drawn Out*. American Institute of Architects, California Council (AIACC). (available at http://www.aiacc.org/2017/11/21/representing-beyond-surface/, retrieved November 2018).

[130] Kanekar, Aarati (2015). Op. cit., p. 117–118.

[131] Corner, James (1999). Op. cit., p. 225.

enacted. The map becomes then a translation of visible and invisible aspects of the landscape, revealing potentialities rather than conforming to its description. In this sense, the map maker is dealing with concrete and abstract aspects of the landscape in a 'dense' way, using layering and collage techniques that enhance its complexity. The density of maps results from the use of representative and interpretive techniques that are in service of the production of a deeper knowledge of the space. Back in 1996, Christian Jacob proposed two theoretical categories of maps—'transparent maps' and 'opaque maps'. The transparent map corresponds to a tradition in the history of cartography that understands maps as devices for the representation of objective knowledge, as a blank canvas to support geographical information applied to it. The opaque map instead corresponds to a new interpretation of maps that is revealed through understanding maps beyond their representative qualities and is immersed in research that study maps from 'graphic, aesthetic and structural points of view and see in them a complex architecture of signs'.[132] Jacob was then making a statement for a 'dense' history of cartography that does not divorce representation from the cultural and historical contexts of its production.

Deep mapping has been associated with narratives and terms such as 'literary geography' or 'literary cartography'. According to Susan Maher, deep mapping can be characterised as 'mapping a plethora of stories—including fictional ones—in order to capture the deep sense of a place'.[133] The expression gained impulse in the humanities in the early 1990s[134] and has been used to describe mapping exercises triggered with the purpose of in-depth investigation of a particular place guided by emotions, perceptions and personal feelings associated with them.

[132] Jacob, Christian (1996). Toward a cultural history of cartography. *Imago Mundi*, vol. 48, pp. 191–198.

[133] Maher, Susan N. (2014). *Deep Map Country:* Literary Cartography of the Great Plains. University of Nebraska Press.

[134] The American writer William Least Heat-Moon is regarded as have been introduced the term with his book *PrairyErth: A Deep Map* (1991). An intersection of travel writing, historical research and sociological study, the book proposal is to deeply chart Chase County in Kansas, USA. Iain Biggs pointed out that the inspirational origins for deep mapping can be traced back to Thoreau's Walden. Biggs, Iain (2010). 'Deep mapping': A Brief Introduction. In *Mapping Spectral Traces.* Edited by Karen E. Till. Blacksburg: Virginia Tech College of Architecture and Urban Studies, pp. 5–8.

Visual and performance artists engage with deep mapping combining geography with artistic representations and narratives of places. Trevor Harris appropriately suggests the term 'deep geography', encompassing deep mapping, spatial storytelling and spatial narrative.[135] Mapping deeply requires 'getting out in the field [...] and looking hard at stuff', Denis Wood asserts.[136] It is a bodily commitment with the land: one's need to immerse in the territory, to go beyond the surface and into profoundness, to 'dive within'.[137] This implies an intimate journey and each map produced will be unique, because every mapping exercise is a personal experience. When Wood describes his first deep mapping project, carried out with undergraduate landscape architecture students in the early 1980s, he refers to the maps produced as having 'fingerprints all over them'.[138]

The terms and definitions proposed to characterise and mark out such practices implicitly assume a strong connection with artistic forms of mapping: deep maps cannot emerge from scientific cartography. There is a hint of criticism towards *thin* maps that remain on the surface, concerned about the precision and accuracy of representation and 'focused on the material and physical characteristics of landscape and society'.[139] GIS technologies are to blame for a standardisation of maps produced for the specific purposes of depicting spatial data, usually by 'slicing' a single reality rather than 'layering' multiple realities. In these terms lies Iain Biggs' claim for an 'open' deep mapping to withstand the harnesses of disciplinary categorisation.[140] However, the same techniques and technologies developed for the improvement of cartography, in particular digital cartography, have contributed to the enrichment of artistic practices, using them as creative tools.

[135] Harris, Trevor. (2015). Deep geography—deep mapping: spatial storytelling and a sense of place. Bodenhamer, David; Corrigan, John and Harris, Trevor (2015). *Deep maps and spatial narratives*. Bloomington, Indiana: Indiana University Press, pp. 28–53.

[136] Wood, Denis (2015) Mapping deeply. *Humanities*, 4(3): pp. 304–318.

[137] Roberts, Les (2016) (ed.) Deep Mapping and Spatial Anthropology. *Humanities*, 5(1): 5.

[138] Wood, Denis (2015). Op. cit., p. 313.

[139] Harris, Trevor. (2015). Op. cit., p. 30.

[140] Biggs, Iain (2010). Op. cit., p. 6.

The next chapter will present a history, with a particular focus on urban cartographies, of how artists got involved in new ways of mapping, not restricted to describing locations but depicting places that one occupies and inhabits. Bringing attention to geographies 'from the ground', very often explored and experienced by the artists themselves, and to the flaws of representations 'from above', artists found ways to engage with spatial practices and the urban 'lived space' and critically challenge cartographical methods—as will be explored in the next chapter.

CHAPTER 4

A History of Art Maps and Mappings

The chronology of the art map, or the history of the intersection between art and cartography, can be contentious. As explored in the previous chapter, in sixteenth-century Europe, cities became the frequent subject of maps, represented according to a defined tradition of mapmaking that culminated in detailed atlases and collections that brought together artistry and engineering. During the eighteenth and nineteenth centuries, this tradition was replaced by a scientific cartography, and the paths of art and science in maps were again separated. This is a history of maps produced to serve as informational and practical pieces, where the art is recognised in the 'embellishment' of cartographic features and the inventiveness of modes of representation.

Parallel with this tradition other maps were created to express concepts and emotions beyond geographical information, maps that belonged more to the realm of fantasy than to territory artistically interpreted. The examples come from literature, as the maps that illustrated the book *Utopia*, by Thomas More, first published in 1516,[1] or *A Plan of the Road from the City of Destruction to the Celestial City*, a map that appeared in the 1821 edition of *The Pilgrim's Progress*, a Christian allegory originally written in 1687 by John Bunyan. An intriguing example often cited (and reproduced

[1] A very well-known version of the Island of Utopia is a map produced by Abraham Ortelius for the 1595 edition.

G. Lanci, *Art Maps and Cities*,
https://doi.org/10.1007/978-3-031-13306-0_4

in the third issue of the *Internationale Situationniste*[2]) is the *Carte du pays de Tendre* (Map of Tenderness) that appeared in 1654 illustrating a novel written by Madeleine de Scudéry. The map is a metaphor of the stages of sentimental relationship; it depicts a central path linking the 'New friendship' town at the bottom to the 'Tenderness' town on top, surrounded along the way by small villages such as 'Sincerity' and 'Affinity'. The terrain also contains dangerous sites, like the 'Rock of Pride' and the 'Lake of Indifference', and it is margined by the 'Sea of Danger' and the 'Sea of Hostility'. Giuliana Bruno, for instance, considers this map as being the first feminist cartographic expression, since it focuses on personal and intimate connections rather than on functionalistic and militaristic ones that were dominant at the time.[3]

Imaginative, autonomous and unruly maps, which could be set apart of the mainstream traditions of cartography, are not a new phenomenon but certainly became more common in the nineteenth century.[4] However, it was at the beginning of the twentieth century that artists engaged with cartography in a completely original way that laid the foundations to the establishment of what can be called as 'contemporary' map art. For the first time artists were incorporating maps in artworks, or making maps as their own artwork, in the context of avant-garde movements and their critical, political and aesthetic implications. More importantly these first manifestations of map art have roots in a historical period of philosophical and scientific revolution that led to rethinking the relationship between vision and space, as discussed in the previous chapter.

The artist-designed map that confronts conventional cartographic practices, creatively challenging traditional representations of space, political power and dominance over the territory, emerged from within the Surrealism and Dada art movements, as suggested by Denis Wood and

[2] The map is placed besides an aerial view of Amsterdam. *Internationale Situationniste* 3, December 1959, pp. 14–15. See also Bruno, Giuliana (2002). *Atlas of emotion: journeys in art, architecture and film*. New York: Verso.

[3] Bruno, Giuliana (2002). Op. cit., p. 76.

[4] The Cornell University Library assembled a collection of 'Persuasive Cartography' including more than 800 maps produced to 'send a message' using 'allegorical, satirical and pictorial mapping; selective inclusion; unusual use of projections, colour, graphics and text; and intentional deception'. The collection covers six centuries but only 50 maps predate 1800; the earliest example dates from 1491. Available at https://persuasivemaps.library.cornell.edu/.

other scholars.[5] Dadaists, in their constant search for something new, rather than honing a skill to perfection, might be better defined as artists who aimed to free themselves of all conventions and past traditions, to try and find a different path for creation. Dada art reflected the confusion and despair that surrounded World War I with nonsensical ideas, appealing to the anarchic, rebellious and experimental. In the midst of their performances, exhibitions and manifestos, maps were collaged into artworks; examples are Hannah Hoch's *Cut with the Kitchen Knife Dada Through the Last Epoch of Weimar Beer-Belly Culture of Germany* (c. 1919–1920); Raoul Hausmann's *A Bourgeois Precision Brain Incites World Movement* (also known as Dada Triumphs! or Dada Conquers, 1920); and Kurt Schwitters' *The Holy Saddler's Portfolio* (1922). They might be located timidly at the margins of the compositions, quite imperceptible, but their presence marks the beginnings of the displacement of the map from its scholarly grounds to the realm of creativity and imagination.

During the 1920s cartography was further explored in Surrealism, an intellectual and artistic movement 'that explicitly engaged cartography as practice rather than simply the map as image' according to Denis Cosgrove;[6] the most well-known artwork of this period is the surrealist map of the world *Le monde au temps des Surréalistes*, from 1929. There are sparse examples from the 1930s and 1940s; Denis Wood, who compiled an extensive and detailed chronology of the art map in the twentieth century, mentions Max Ernst's *Europe After the Rain I* (1933), Joseph Cornell's earlier boxes with maps in them (1936–1943), Marcel Duchamp's *Allégorie de genre* (1943) and Joaquín Torres-García's upside-down map for the artists collective La Escuela del Sur (1943).[7]

The most significant artistic shift towards mapping cities was observed with the situationists and conceptual artists in the mid-1950s and 1960s. Artists took on cartography as a tool and a medium to experience the physical environment and in particular the urban space, as is the case with the situationists. Producing maps became itself a method to create meaning, through urban explorations such as *dérives*, and trigger

[5] Cosgrove, Denis (2005). Maps, mapping, modernity: art and cartography in the twentieth century. *Imago Mundi*, 57(1), 35–54; D'Ignazio, Catherine (2009). Art and Cartography. In Kitchin R, Thrift N (eds) *International Encyclopaedia of Human Geography*, Volume 1, pp. 190–206. Oxford: Elsevier; Wood, Denis (2010). *Rethinking the power of maps*. New York: The Guilford Press.

[6] Cosgrove, Dennis (2005). Op. cit., p. 39.

[7] Wood, Denis (2006). Map art. *Cartographic Perspectives*, 53, 5–14.

transformation through political statements. Since then, mapping has been used as a technique to explore urban spaces bringing the unseen to visibility, being either the fantastic or surreal, composing improbable landscapes, or to unveil untold stories of urban reality. Through the 1980s the art map concept gained the status of a genre, and by the end of the 1990s, there was a considerable collection of exhibitions and publications dedicated to the theme.[8] At the turn of the twenty-first century, the relationship between art and cartography 'reached another turning point with the growing interest in approaching artwork as a source of spatial information that could be mapped'.[9] The representation of the world, as well as its perception, was again challenged. Abstracting reality became a more complex and less defined task, as the notion of reality proper was transformed by the advent of digital technologies. Once more, artists 'intruded' into the realm of cartography manipulating its tools to capture new spatial discourses and practices, as we will see later in this chapter.

Walking and Mapping: From Drifting to Dissecting (1910s–1970s)

In his book *Walkscapes: Walking as an Aesthetic Practice*, Francesco Careri identifies in art history three "moments of passage" when walking intrudes the art world in a significant manner: from Dada to Surrealism (1921–1924), from the Letterist International to the Situationist International (1956–1957) and from Minimal Art to Land Art (1966–1967).[10] The artists involved in those art movements were interested in taking artworks outside the art gallery, deconstructing processes and spaces of art production, redefining art as experience, incorporating art in the political sphere and repudiating art as an object-oriented practice. In doing so artists took their works and themselves to unusual non-arty venues such as bars or cabarets and, above all, to the streets. What these artists also have in common is the incorporation of mapping into their practices while reclaiming new approaches in relation to life in cities.

[8] Watson, Ruth (2009) Mapping and contemporary art. *The Cartographic Journal*, 46 (4), pp. 293–307.

[9] Ribeiro, D. M, and Caquard, Sebastian. Cartography and Art. Wilson, John P. (ed) (2018). *The Geographic Information Science & Technology Body of Knowledge*. DOI: https://doi.org/10.22224/gistbok/2018.1.4.

[10] Careri, Francesco (2002). *Walkscapes: walking as an aesthetic practice*. Barcelona; Gustavo Gilli, p. 21.

Before Dada and Surrealism, predecessors of the urban wandering can be found in the early nineteenth-century Europe. Making use of literary expressions to describe the transformation of capital cities on the continent, Thomas de Quincey in the 1820s, strolling around London, fuelled by opium, and Charles Baudelaire drifting in Paris in the 1850s, under the effect of intoxicants, are prototypes of the traditional figure of the flâneur. Their self-indulged, solitary experiences are narrated in a lyric style, used to express very personal emotions and feelings and to enhance their discoveries around the city, as exemplified in this extract from de Quincey's *Confessions of an English Opium-Eater*:

> Some of these rambles led me to great distances, for an opium-eater is too happy to observe the motion of time; and sometimes in my attempts to steer homewards, upon nautical principles, by fixing my eye on the pole-star, and seeking ambitiously for a north-west passage, instead of circumnavigating all the capes and head-lands I had doubled in my outward voyage, I came suddenly upon such knotty problems of alleys, such enigmatical entries, and such sphynx's riddles of streets without thoroughfares, as must, I conceive, baffle the audacity of porters and confound the intellects of hackney-coachmen. I could almost have believed at times that I must be the first discoverer of some of these *terræ incognitæ* and doubted whether they had yet been laid down in the modern charts of London.[11]

Quince's prose is replete with spatial references and allusions to movement and travel, positioning himself as an explorer of the urban environment. One generation later a more objective kind of exploration was conducted by the Franco-Peruvian socialist philosopher and activist Flora Tristan, who in her writings conceives a *flâneuse* who is both a social investigator of the city and an intimate observer of public life.[12] Although Tristan's promenades and peregrinations were purposefully committed to her social and feminist causes, in a different approach to de Quincey's travelogue prose or Baudelaire's urban poetics, she poses herself as a 'pariah' and a 'foreigner' in the modern metropolis, offering

[11] De Quincey, Thomas (2008). The pleasures of opium. *Confessions of an English opium-eater*. Cambridge: ProQuest LLC/ Literature Online Prose. (Digital edition).

[12] Nesci, Catherine (2001). Flora Tristan's urban odyssey; notes on the missing flâneuse and her city. *Journal of Urban History*, vol. 27, no. 6, pp. 709–722. See also Wolff, Janet (1985). The invisible flâneuse: women and the literature of modernity. *Theory, Culture & Society* 2, pp. 37–46.

a particular gaze on the modern metropolis, writing about urban spaces 'from the margins'.[13]

These first literary and sociological expressions of urban wandering directly influenced the visual expressions of Dada and Surrealism art movements, which in turn inspired situationists, land artists and conceptual artists. Their historical context, across the turbulent years between the 1920s and 1960s, comprised the Second World War and many other conflicts over territorial boundaries, battles for independence and fights for identity recognition. Politics of space and cultural appropriation were themes not ignored by artists drawn into the urban environment as a locus for their practices, looking at cities as privileged spaces for roaming and creating, but also protesting. Walking and wandering became aesthetic acts that explored, among other things, new dimensions of urban spaces. These walking acts were also related to a form of ***anti-art***, a term that was coined by Marcel Duchamp in 1913,[14] an artist whose association spanned from Cubism in the early twentieth century to Conceptual Art in the 1960s.

Careri, attempting to distinguish the various ramifications of the urban wandering as art, considers that dadaists were in search of a 'banal city' away from the bourgeois city; the Surrealists went beyond pursuing the 'oneiric city', exploring the territory of fantasy and of the unconsciousness; and the situationists after them explored the 'playful' and nomadic' city, rejecting personal representations and advocating a collective and revolutionary art combined with the practice of walking.[15]

Dadaists organised 'visit-excursions' to places away from the tourist-beaten tracks of Paris. According to Careri 'this was the first time that art rejected its assigned places, setting out to reclaim urban space'.[16] The ***déambulation*** (perambulation) Dadaist would be later recovered by the Letterist movement with its theory of drifting, but it would be the situationists in the 1950s who would take on cities as their main mapping

[13] Nesci, Catherine (2001). Op. cit., p. 720.

[14] Other previous experiments can be related to the concept of 'anti-art' as defined by Duchamp. Around 1882 Jules Levy founded in France ***les arts incohérents*** (the incoherent arts), a movement that presented a playful side of the art world, mocking art masterpieces (such as *The Joconde smoking pipe*, by Sapeck, a defaced version of Mona Lisa) and producing work against the art canons, being irrational, nonsensical and very often humorous. Although seen as eccentricity that was short lived (1882–1891) and confined to a small circle of friends, there are many characteristics there that could be related to Dada and surrealist artists.

[15] Careri, Francesco (2002). Op. cit., pp. 85–86.

[16] Careri, Francesco (2002). Op. cit., p. 21.

subject. Initiating a fertile and creative period in artistic cartography, this could be considered a 'watershed' between the more traditional, representative urban mapping, one that intends to 'mimic' its visual features with no other implications in social and cultural dimensions and a more contesting, interpretative urban mapping that attempts to be pluri-dimensional. The maps produced by the situationists inaugurated a phase of provocative, critical and radical mapping practices that later on in the twenty-first century would inspire new and even more disruptive art movements.

The Situationist International: Dérives, Psychogeography and Unitary Urbanism

The Situationist International (hereafter SI), founded in 1957, was a Paris-based collective of visual artists, writers and thinkers. As fierce critics of modern capitalism, urban planning and consumerism, the situationists engaged in writing, artistic performance and manifestos that made them one of the most prominent groups of the mid-twentieth century European artistic and political avant-garde.

Surrealism and Dada were major influences for the situationists, who also directly incorporated the Lettrism of Isidore Isou in the mid-1940s, the CoBrA movement of late 1940s and the Letterist International, whose existence between 1952 and 1957 paved the way for the foundations of the SI. The SI also amalgamated the International Movement for an Imaginist Bauhaus, an offshoot of CoBrA, and the London Psychogeographical Association, founded by British artist Ralph Rumney. Guy Debord, a filmmaker, writer and political activist, was the central figure of both the Letterist International and the Situationist International. Other artists, like painters Asger Jorn and Constant Nieuwenhuys, both founding members of the CoBrA group, became some of the main collaborators of Guy Debord in the initial years of the SI, a period that Tom McDonough called their 'architectural interlude', when Constant proposed the New Babylon project for a future city incorporating situationists' ideas, as will be commented later in this chapter.[17]

The situationists were ambitious. As Guy Debord later wrote in 1972, 'It is known that initially the situationists wanted at the very least to build

[17] Constant left the group in 1960 and after 1962 the movement entered a period of 'more abstract and rigid theorizing'. McDonough, Tom (2009). *The situationists and the city*. London: Verso, p. 2.

cities, the environment suitable to the unlimited deployment of new passions. But of course, this was not easy and so we found ourselves forced to do much more'.[18] As a movement, the SI did not present itself as political, but critique of consumerism and urban life was at the centre of its debates and eventually they played a predominant role in the student revolts of May 1968 in Paris. The group was dismantled in 1972 but the ideas disseminated by the situationists remain until today a lively subject of academic essays and books, which production is helped by a significant legacy of posters, booklets, manifestos and other printed materials available for research.[19]

The construction of situations was a main artistic and philosophical practice that supported the theories of situationists. Change, revolution and non-conformism were sought to guide the creation of situations. As defined by Guy Debord situations should be moments of life 'concretely and deliberately constructed by the collective organisation' to lead to a 'unitary ambiance and a play of events'.[20] The city was the stage for situations, where art could be experienced continuously, being part of daily life, and in this sense, they intended to move away from traditional places of art production (the artist's studio) and art exhibition (the art gallery/museum). This 'art on the streets' was unpredictable and spontaneous. Situationists wanted to expose the appeal to consumerism in the materialistic society by devaluing objects and institutions that promoted them, including here the 'official' art exposed in museums and galleries and the social and economic system that supported it. Art for them should be released from these obstacles and lived on the streets, continuously.

[18] Debord, Guy. On wild architecture. Preface to Ezio Gribaudo and Alberico Sala's *Jorn / Le jardin d'Albisola* (1972). Available at http://www.cddc.vt.edu/sionline/postsi/wild.html.

[19] The activities and thoughts of the SI members are well documented on the *Internationale Situationniste* journals. Between July 1958 and September 1969, the group published 12 issues of them, the first comprising 32 pages, the last one 116 pages. A digital edition of all issues (in French) made of black and white photocopies can be found here http://www.larevuedesressources.org/internationale-situationniste-integrale-des-12-numeros-de-la-revue-parus-entre-1958-et,2548.html. For an English translation of the main articles, see Ken Knabb's website http://www.bopsecrets.org/SI/.

[20] In 'Definitions' (1958) in *Situationist International anthology*. Ed. and trans. Ken Knabb (Berkeley, Bureau of Public Secrets, 1981). Available online http://www.bopsecrets.org/SI/1.definitions.htm (revised and expanded edition, 2006) and at http://www.cddc.vt.edu/sionline/index.html (Situationist International Online).

In many of their critical texts, the situationists denounced the apathy imposed by modern life, its causes related to an alienated relation with the urban environment, and the politics of space, as a result of the dominant mainstream media. Passivity and doubt should be replaced by activism, play and instigation; therefore, the creation of situations was the vehicle for performing ideas. Refusing repetition and mesmerism were key to construct situations that could actively transform everyday life. These ideas were first conceived in Ivan Chtcheglov's *Formulary for a New Urbanism*, published in 1953, where the term *dérive* is firstly used within the situationist framework. This is a fundamental document of the Letterist movement and a main contributor to the concepts and practices that would later define the SI. Chtcheglov begins his essay remarking on the boredom experienced in the city, deploring the absence of imagination and narrative in urban experience and denouncing an existence of dullness to what people are condemned in the functional world of modern urbanism:

> A mental disease has swept the planet: banalization. Everyone is hypnotized by production and conveniences — sewage systems, elevators, bathrooms, washing machines.[21]

In order to overcome this state of indifference, Chtcheglov calls for 'a modifiable built environment that might change with the desires of its inhabitants [...] a city whose quarters might correspond to differing emotional states'.[22] In such cities people could dedicate themselves to infinite wandering:

> The main activity of the inhabitants will be CONTINUOUS DRIFTING: the changing of landscapes from one hour to the next will result in total disorientation.[23]

'Total disorientation' here could be understood as another provocation to combat the callousness of the modern city. Situationists called for

[21] Chtcheglov, Ivain (1953). *Formulary for a New Urbanism*. Available at http://www.bopsecrets.org/SI/Chtcheglov.htm.

[22] McDonough, Tom (ed.) *The Situationist International and the city*. London: Verso, pp. 32–33.

[23] Chtcheglov, Ivain (1953). Op. cit. 'Continuous drifting' (*dérive* continue) is in uppercase in the original text. See http://www.larevuedesressources.org/IMG/pdf/internationale_situationniste_1.pdf (p. 19) (this is an abridged version published in the Internationale Situationniste n. 1).

pro-active interaction with the city, to engage with it rather than passively consume what is given by planners and administrators. The dérive is proposed as a tool to instigate creative engagement but also to transgress the norms of organised living: in this sense, the pleasure gives way to a critical practice.

Art making and theorisation were undertaken as subversive practices,[24] aiming to undermine the so-called society of spectacle and the fragmentation of everyday life where entertainment is passively consumed. Graffiti, comics and collage were instrumental to subvert traditional means of language, and the situationists made good use of them in their regular publications, as their periodical *Internationale Situationniste* (1958–1969) and the books *Memoires* and *Fin de Copenhague* (both published in 1957) authored by Guy Debord and Asger Jorn, featuring collages of newspapers, comics, magazines, photographs, advertisements and cartoons. Modern urbanism and its attempt to rationalise and homogenise urban life was another target of strong criticism. The situationists understood that precision would kill diversity and spectacle would commodify the joys of freely exploring and living in cities. They advocated that architecture should be designed according to human behaviour. In 1962 Constant proposed the *New Babylon* project, inspired by Unitary Urbanism and its idea that 'the city is a point of departure for free play', understanding cities as 'models for thought and recreation of the future'.[25] Cities, according to Constant, should be designed for a new way of life in an automated world, assuming that human beings would be freed of mechanical labour and repetitive tasks, and then executed by robots and other automated engines. Humans should be free to engage in creative work, and New Babylon was the archetype of a playful city for the (future) playful human.

Roaming, wandering and traversing the urban space freely, with no clear destination, objective or timeframe has been explored in artistic practice since the nineteenth century, as already mentioned with the writings of the French poet Charles Baudelaire. As a point of departure to experience the city, *flânerie* has proved to be a recurrent method in art to apprehend, document and express subjective perceptions of the public urban

[24] Around the theme of 'subverting cartography', Pinder considers that the SI maps could not be considered exactly 'subversive' per se, but instead they helped to 'foster a subversive attitude to cartography'. Pinder proposes that a subversive practice cannot be something 'absolute, singular' but rather involve a 'continual process that constantly turns against itself'. Pinder, David (1996) Subverting cartography: the situationists and maps of the city. *Environment and Planning A*, volume 28, pp. 405–427 (p. 424).

[25] Chtcheglov, Ivain (1953). Op. cit.

space in the industrial and post-industrial eras. In the early twentieth century the metropolis was the epitome of the modern city: an ever-changeable landscape of simultaneous events. Futurists and dadaists were the first artists to embrace the rapid urban transformation in Western European cities, also incorporating public urban performances in their practices.[26] On moving around, choosing venues and performing, these artists were taking geography into account and maps would soon start to be used in more explicit ways.

The situationists were interested in surveying the urban space as much as they wanted to register their experiences. Maps were widely used to illustrate urban drift and psychogeography in practice. The *dérive* was a new approach to urban surveying and so was the situationist map a new medium to represent it. The *dérive* (translated into English as drifting) and *détournement* (deviation, deflection) were used on excursions as tools and a technique of random and rapid passage through varied neighbourhoods in the city that allowed a more creative and sensory approach to urban surveying. In *Introduction to a Critique of Urban Geography* (1955), Debord wrote that a 'renovated cartography' should be used as means of 'intervention' and psychogeographical maps could:

> [...] contribute to clarifying certain wanderings that express not subordination to randomness but total *insubordination* to habitual influences (influences generally categorized as tourism, that popular drug as repugnant as sports or buying on credit).[27]

Insubordination is the key word here, and the mention of tourism reflects the idea that the map should not be reduced to a guide or a crutch to thoughtless excursions in the urban space. The environment of experiment for the SI was Paris, and mapping it through the techniques of dérive and détournement became a regular practice. One of the first examples is the *Axis of Exploration and Failure in the Search of a Situationist 'Great Passage'*[28] made in 1957. The map brings cartography and art together in

[26] Hollevoet, Christel (1992). Wandering in the city: flânerie to dérive and after: the cognitive map of urban space. *The power of the city/ the city of power*. Exhibition catalogue. Whitney Museum of American Art, pp. 25–27.

[27] Debord, Guy (1955) *Introduction to a Critique of Urban Geography*. Transl. Ken Knabb (2006). Available at http://www.cddc.vt.edu/sionline/presitu/geography.html.

[28] Great Passage here refers to the northwest route to East sought by European explorers since the fifteenth century.

a collage of distinctive elements: a photo of Debord and Chtcheglov (top left), cut outs of aerial photography of sites in the left bank of the Seine (in the middle) and a reproduction of an evocative painting [29] (bottom right). Subverting the meaning of the map as a document to impose order and rationality was enhanced in the situationist discourse. As Wark puts it, 'Debord was a sort of street ethnographer, although his method was more intoxicant peregrination than participant observation'.[30]

Psychogeographical maps should open ways to alternative realities, often leading to personal and unique interpretations rather than representing universal truths. The map is thus a register of a subjective experience with no commitments to precision and science:

> Beyond the discovery of unities of ambiance [...] one comes to perceive their principal axes of passage, their exits and their defences. One arrives at the central hypothesis of the existence of psychogeographical pivotal points. [...] With the aid of old maps, aerial photographs and experimental dérives, 'one can draw up hitherto lacking maps of influences, maps whose inevitable imprecision at this early stage is no worse than that of the first navigational charts. The only difference is that it is no longer a matter of precisely delineating stable continents, but of changing architecture and urbanism.[31]

Guy Debord and Asger Jorn produced two psychogeographical maps of Paris: the *Guide Psychogeographique de Paris: discours sur les passions de l'amour* and *Naked City* were published in 1957.[32] These maps described

[29] *Seaport with the Embarkation of Saint Ursula.* Claude Lorrain, 1641(National Gallery, London).

[30] Wark, Mckenzie (2011). The beach beneath the street: the everyday life and glorious times of the Situationist International. London: Verso, p. 16.

[31] Debord, Guy (1956). Theory of the Dérive. *Les Lèvres Nues* #9 (November) reprinted in *Internationale Situationniste* #2 (December 1958). Trans. Ken Knabb. Available at http://www.cddc.vt.edu/sionline/si/theory.html.

[32] According to Gilles Rion, *The Psychogeographical Guide of Paris* was part of a selection of five maps drawn up by Guy Debord for the 'First Exhibition of Psychogeography' at the Taptoe Gallery in Brussels in February 1957. He adds that 'after Debord's refusal to take part in this collective exhibition (which also brought together Asger Jorn, Yves Klein, Ralphy Rumney, Michèl Bernstein and Mohammed Dahou), this collage was finally printed with *The Naked City* in May, during a visit to Denmark with Jorn, and published by the International Movement for an Imaginist Bauhaus. Both were then reproduced in 1958 in Jorn's book *Pour la forme*, published in Paris by the Internationale Situationniste'. See http://www.frac-centre.fr/_en/art-and-architecture-collection/debord-guy/guide-psychogeographique-paris-discours-sur-les-passions-l-amour-317.html?authID=53&ensembleID=135.

the unconventional experiments carried out in Paris and are both collages made of cut outs of then easily available printed maps: Blondel la Rougery's *Plan de Paris a vol d'oiseau*, 1953, was the source for the *Guide psychogeographique*, while *Naked City* was sourced from the *Guide Taride de Paris*, 1951.[33] At first sight they may not appear so alternative or even innovative; however, on defragmenting spaces and deconstructing itineraries, the psychogeographic maps were confronting the logical structure of traditional maps. Paris is presented as a collage, an assemblage of parts that is united through routes; the city is given meaning as it is explored on foot. The map registers a psychological state of ambulation when creativity is grasped to inspire alternative perceptions of the urban space.

The situationists frequently used aerial photography; in their publication *New Theatre of Operations in Culture* (1958), there is a reproduction of a mosaic of aerial photographs depicting south-eastern Paris. Under the photograph a diagram shows the interrelation of practices and concepts for a situationist experimentation: continuous play activity, dérive, psychogeography, experimental behaviour, détournement, unitarian urbanism and construction of situations. The use of aerial photographs was mostly inspired by a publication of 1952, *Paris and the Parisian Region*, by the urban sociologist Paul-Henry Chombart de Lauwe, himself an enthusiast of aerial photography and demographics.[34] On the one hand, on using aerial photography and mapping, there is an attempt by the situationists to adopt a 'universal' international language, modernising the surrealist discourse that was at the genesis of the group and moving towards a more holistic, integrated perception of the urban landscape; on the other hand, the situationist map draws attention to the fragmentation and codification of the urban space imposed by systems of representation that exclude people and their practices in the everyday urban life, such as their perambulations around the city (and therefore the importance on marking routes in contrasting colours in their maps).

The mapping impulse of the first years of the SI began to wane around 1962 as members of the artistic avant-garde associated with the group at its foundation, among them Asger Jorn, split off to form the 'Second

[33] The map fragments of *Naked City* can be seen here overlaid onto Google map https://isinenglish.com/2016/01/09/debords-the-naked-city-mapped-onto-google-maps/.

[34] McDonough, Tom (2009). Op. cit., p. 14. Chombart de Lauwe had published in 1948 'La Decouverte Aerienne du Monde', where he makes use of the recent technologies for land surveying developed during the WW II.

Situationist International',[35] and the political agenda took over the activities of the remaining members congregated around Guy Debord. Meanwhile, first in the USA and later elsewhere, in Western Europe, East Asia and South America, artists were occupied in questioning the very nature of art and art practices. Concerned to move away from the art object, refusing commodification and challenging traditional places of art making and art exhibition, these artists focused on experimental, interdisciplinary and performative art, which also included mapping, as will be discussed in the next section.

After the SI: Fluxus, Conceptual Art and Land Art

In the late 1950s, around the same time of the foundation of the SI in Paris, new art movements were emerging in New York that would impact the art world globally. Loosely following on the footsteps of the European avant-gardes of the early twentieth century, rebelling against the art status quo and embracing counterculture, visual and performance artists initiated a period of radical experimentation that incorporated 1940s' abstract expressionism and developed into Neo-Dada, Pop art, assemblage and happenings. A few art maps were produced then by artists, such as Claes Oldenburg (*Map of Chicago Stuffed with Soft Numbers*, 1963), but probably the most well-known and widely exhibited is a series of paintings titled *Map* that Jasper Johns produced between 1960 and 1967, based on maps of the USA. However, the 1960s brought other developments in the art world that were far more intriguing, including some mapping experiments, namely, the Fluxus group, the conceptual art movement and the land artists.

Fluxus was a creation of the Lithuanian American artist and architect George Maciunas, influenced by the experimental music of John Cage and Eastern philosophy. As indicated in the name, the group was interested in the flow of producing and experimenting with art, paying more attention to the process rather than the product. From its very beginning the group was characterised by an international and multidisciplinary association of musicians, visual artists and designers. Their first collective event was staged in 1961 in New York, and soon after they spread their activities to Germany and Japan. Some of the Fluxus artists engaged in the creation of innovative artworks that combined documentation, spatial narrative and performance, such as Yoko Ono (*Map Piece*), Nam June Paik (*FLUXUS*

[35] Wark, Mckenzie (2011). Op. cit., pp. 104–105.

Island in Décollage OCEAN, 1963) and Maciunas himself, who produced posters and diagrams using maps for documenting Fluxus events and activities. He also collaborated with artist Mieko Shiomi in the series *Spatial Poems*, and together they produced the *Fluxusatlas*, a collection of maps of the world recording Mieko's mail-art events realised between 1965 and 1975.[36]

Yoko Ono's *Map Piece* is part of her book *Grapefruit*, originally published in 1964. The book contains a set of 'pieces', in tandem with the Fluxus' concept of 'event scores', scripts that describes actions to be performed to create the artwork. The *Map Piece* thus creates a map that comes into existence by the actions of whoever follows the set of instructions provided by the artist (including the artist her/himself). In Ono's concept the map is drawn within an urban setting, as indicated by references to 'street' and 'city'. *Map Piece* also implies social interaction and chance encounters that could be typically enhanced in a city environment.[37]

> MAP PIECE
> Draw an imaginary map. Put a goal mark on the map where you want to go. Go walking on an actual street according to your map. If there is no street where it should be according to the map, make one by putting the obstacles aside. When you reach the goal, ask the name of the city and give flowers to the first person you meet. The map must be followed exactly, or the event has to be dropped altogether. Ask your friends to write maps. Give your friends maps.
> 1962 summer

In the summer of 1967, the same year that Guy Debord published his seminal work *La société du spectacle*,[38] an article titled 'Paragraphs of Conceptual Art' appeared in the *Artforum* magazine, signed by American artist Sol LeWitt. In there, LeWitt stated that it was the 'idea' or 'concept' that is the prominent aspect of the artwork; 'the idea becomes the machine that makes art', as he phrases it.[39] The first use of the term 'conceptual' in

[36] Shimoi poems were requests to the public sent via post, and the responses received were charted in maps. Wollen, Peter. Mappings: situationists and/or conceptualists. Newman, Michael and Bird, John (eds) (1999). *Rewriting Conceptual Art*, London: Reaktion.

[37] In 1964, but the time of the book's publication, Yoko Ono produced a second *Map Piece*, where it can be read: 'Draw a map and get lost. 1964 Spring'.

[38] *The society of spectacle*, first translated into English in 1970 by Fredy Perlman.

[39] LeWitt, Sol. Paragraphs in Conceptual Art. In Harrison, Charles and Wood, Paul (2003). *Art in theory, 1900–2000: an anthology of changing ideas*. Oxford: Blackwell.

the art context is attributed to Ed Kienholz in the late 1950s and to Fluxus artist Henry Flynt, who wrote about it in 1961, but it was not until LeWitt's article that conceptual art entered in the mainstream art world, in a succession of exhibitions and publications.[40]

The roots of Conceptual art could already be detected in Marcel Duchamp's *anti-art* and the *ready-mades* he produced in the 1910s. Duchamp's approach to the aesthetics of the artwork through irony and self-criticism would be rediscovered and reinvented decades later 'not just another particular kind of art [...] but an attempt at a fundamental redefinition of art as such', as philosopher Peter Osborne observed.[41] Marcel Duchamp, who was also at the forefront of Surrealism, embraced conceptual art in the 1960s as opposed to a purely 'retinal art'—his expression—that intended to only please the eye. In his own definition conceptual art was 'in the service of the mind'.[42] Minimalists, conceptual artists and land artists embraced the idea of depleting the art object of its 'aura', being demystified, displaced and made incongruous. Without the artefact the visual elements of this 'dematerialised' art could be experienced through interventions and registers made by the artist—text, photographs and maps—that would function as supports for documenting, exposing and *re-experiencing* the artwork.

As a philosophical movement Conceptual art can be placed against an aesthetic theory of the arts (that emerged in Renaissance naturalism and the visual paradigm of the Enlightenment that went through Kant and nineteenth-century aestheticism) and allied with a tradition that goes from Romanticism, through Hegel, Duchamp, Surrealism and the close predecessors Fluxus and Minimalism art movements.[43] Conceptual art has been defined as 'post-object art', 'art-as-idea', 'theoretical art' and other terms that attempt to encapsulate the diversity of practices and art forms that were put under a wide umbrella. Artists linked to the movement positioned themselves as controversial and unorthodox practitioners, who adopted ideas or concepts as a medium to produce art and challenged institutional art categories proposing new forms of art interaction through

[40] Newman, M. and Bird, J. (eds) (1999). Op. cit., p. 5.

[41] Osborne, Peter (1999). Conceptual art and/as philosophy. Newman, M. & Bird, J. (eds.). *Rewriting Conceptual Art,* London: Reaktion, p. 48.

[42] Marcel Duchamp as quoted in Arnason, Harvard H. and Prather, Marla F. (1998). *History of Modern Art: Painting, Sculpture, Architecture, Photography.* 4th edn. New York: Harry N. Abrams, p. 274.

[43] Osborne, Peter (2013). Op. cit., p. 46. See also Osborne, Peter (1999). Op. cit., p. 50.

happenings, performances and site-specific installations. They shared a fascination with maps and mappings, 'not only as a form of documentation but also as a form of design', as observed by film theorist Peter Wollen.[44] However, while the situationists used maps as instruments for their critique of the standardised city planning schemes for the modernisation of Paris, conceptual artists were less concerned with local urban militancy and focused on mapping as another way to experiment with performance within an urban context. The style of politics inside the conceptual art movement was primarily directed to the art world itself, even though many artists engaged in civil rights movements and similar political causes.

Wollen observes that in conceptual art maps 'there is a kind of scientificity, an almost clinical mind-set, based on an obsession with theoretical methodologies for documenting behaviour'.[45] One of the best examples is the *I Went* series, produced by Japanese artist On Kawara. Starting in 1968 *I Went* is an elaborate personal diary that has the urban as a background for his own experimentation; it is not about a particular city but about his movements in various urban settings. Neither are the artist's maps evocative of *dérives*; instead they are precise pieces of documentation, meticulously organised. For the 12 years that comprised the project, 12 volumes were produced, totalling a staggering number of 4740 pages, each containing the itinerary of one day in On Kawara's life. On Kawara's journeys are traced in red lines over black and white reproductions of city maps, which are zoomed in or out to fit in the whole extent of the itinerary in the paper size. The cities look all the same in the background photocopies; they only come into existence by the trace of the artist. The maps themselves are not the central piece of his artwork, or its ultimate goal, but a platform for the mapping with an artistic intent. Each map is thus the registration of such intents—or performances—conducted and programmed by the artist. In this sense he confronts the idea of cartography as representations of places, demystifying its powers of geographical accuracy.

In 1968, the same year that On Kawara launched his mapping diaries, Douglas Huebler, an American conceptual artist, inaugurated his *Site Sculpture Project* with *42nd Parallel Piece*. Huebler's maps questioned the nature of art itself, to begin with his redefinition of the term 'sculpture' and playing with cartographic concepts purely to create drawings.

[44] Wollen, Peter (1999). Op. cit., p. 29.
[45] Wollen, Peter (1999). Op. cit., p. 36.

Following this concept, in *42nd Parallel* he traces a line over an ordinary map that contains the location of 14 towns 'either exactly or approximately on the 42° parallel in the United States' that were marked 'by the exchange of certified postal receipts sent from and returned to "A"—Truro, Massachusetts', according to the description inserted just below the title of the map. In another work, *Variable Piece 1*, he draws concentric squares over the street grid of Manhattan, marking the corners as 'A', 'B', 'C' and 'D'. Huebler's maps are the result of strategies to investigate the art process, and mapping is a method to deconstruct visual paradigms and representational systems, imprinting a new meaning to the artwork. His maps, like On Kawara's cartographic diaries, are material evidence of performative acts rather than mere illustrations or aesthetic objects. In the case of Huebler he went further in assembling maps, text and photography into pieces conveying a full concept of his experiences, in space and time, that could be also reconstructed by the viewer,[46] resembling Yoko Ono's intentions to give to the observer/reader/walker the conditions to perform an artistic piece.

The possibilities for recreating experiences through artworks were equally explored by land artists, also called 'environmental' or 'earth' artists, in direct, large-scale interventions into spaces, usually outdoor settings, transforming their aspect and configuration. Landscapes were the primary object for observation and a medium to investigate 'earth' materials, such as soil, rocks and clay, to recreate, redesign and imprint marks and structures, using the territory as a malleable canvas. Although most of the land art is performed and placed in non-urban sites, its concept of modifying the land, and so recreating the map, resounds in other practices that have the city as a theme.

The movement emerged in New York, where the first exhibition entitled 'Earthworks' took place in 1968 at the Dwan Gallery. Robert Smithson was a prominent figure on the American scene, as practitioner and critic, and a pioneer of an art form made *in* the earth and *from* the earth: 'I'm more interested in the terrain dictating the condition of the art', said Smithson.[47] His series of *Non-sites*, first exhibited at the Dawn

[46] Huebler considered his performances as 'programmatic journeys in the real world'; using maps he was creating 'a new frame of reference for art'. Wollen, Peter (1999). Op. cit., p. 40.

[47] Flam, Jack (1996). *Robert Smithson: the collected writings*. Berkeley: University of California.

Gallery, presented materials collected from outdoor locations and arranged in crafted boxes, together with information about the geography and geology of the areas from where the material had been removed. Smithson had been working with the concept of the artwork as a relationship between outdoors ('sites') and indoors ('non-sites'), which he called a dialectic of the 'raw reality' of the site against the 'abstract container' of the non-site.[48] In other words, the non-site is space abstracted inside a double container—the boxes, maps and photographs that replicate the site, set up in the gallery 'box'—while the site is space 'lived', to use Lefebvre's term, where the 'raw' art/performance took place.

Following this Smithson also made 'earth maps', drawn, performed and constructed with materials found in the site and then left on the site. These are artworks that most characterise 'land art', and in the case of Smithson, his most well-known piece was *Spiral Jetty* (1970), a sculpture made of rock, mud and salt crystals, shaped as a coil that projects towards the waterline and can be walked via a 460-meter long path, built on the shore of the Great Salt Lake, USA. Smithson produced a video documenting its construction, as mapping of the whole process, but *Spiral Jetty* is a map in itself, an inscription on the surface that becomes the land, 'literally sinks into it' and turns into a 'spatiotemporal embodiment of a journey on the earth's surface'.[49]

In Britain, maps and mapping were also an integral component of the practice of the land artist Richard Long who, like Smithson, brought the outdoors into the gallery space. One of his first works is *A Square of Ground* (1966), a three-dimensional model of a landscape cut out as a map grid. Although not made of collected material, but composed in the art studio, with plaster and paint, Long's piece suggests re-connection with the land through re-enacting the experience of the viewer hovering above it. Long's main mapping practised early on had developed from his walks, as in *A Line Made by Walking*, an artwork produced in 1967 that made him best known as minimalist and conceptual artist. In later works Long explored further the possibilities of combining performance, visual documentation and text, presented as art installations of 'walk sculptures'.

[48] Flam, Jack (1996). Op. cit., p. 178.

[49] In fact the sculpture sunk into the site, to resurge almost 30 years later. Casey, Edward. (2005). *Earth-Mapping: artists reshaping landscape*. University of Minnesota Press. p. 19; 24.

Mapping, Counter-mapping and the Art Map Age (1970s–2000s)

In parallel with the conceptual and land art movements, intellectuals engaged in debates around society and territories were shifting their attention to cartography aiming to better understand relations of power. As examined in the previous chapter, traditional cartographies, incorporating formulaic and descriptive mapping practices, were challenged in academia at the same time as new mapping technologies were establishing more accurate and scientific ways to map the world. In his book *Rethinking the Power of Maps*,[50] an updated version of *The Power of Maps*, first published in 1992, Denis Wood established a simple division in mapping practices: 'mapping' and 'counter-mapping'. This division is also chronological, where the first group includes mapping practices that serve the interests of the state, corresponding to a history of around 600 years, and in the second group Wood includes those that resist the power of the state, emerging in the middle twentieth century.[51] In this section the focus is on how such criticism percolated into the art world.

One of the earliest examples of interchanges between scholarly work and artistic practices is the work of the geographer William Bunge who, in the 1960s and 1970s, mapped disadvantaged neighbourhoods in Detroit. His work is representative of what later was recognised as 'counter-mapping' or 'radical cartography', which in turn inspired artists in practices that combined aesthetic defiance and activism.[52] William Bunge founded the *Detroit Geographical Expedition*, and in 1971, they produced one map that summarises in a sharp, irrevocable way the state of social relations in Detroit at that time. It is a very simply designed map, as in early computer-based cartography, black and white, showing the network of streets as single lines, a shade of grey depicting residential areas and points marking the locations of the events, in this case, accidents involving automobiles and pedestrians. What is not so simple is to confront its straightforward title, 'Where Commuters Run Over Black Children on the Pointes-Downtown Track', that defiantly gives an interpretation of the data collected. In a more mannered and scientific fashion the map is indicating a

[50] Wood, Denis; with John Fels and John Krygier (2010). *Rethinking the Power of Maps.* New York, the Gilford Press.

[51] Wood, Denis (2010). Op. cit., p. 7.

[52] Mogel, Lize and Bhagat, Alexis (eds.) (2008). *An Atlas of Radical Cartography,* Los Angeles, Journal of Aesthetics & Protest Press.

geographical pattern of events, but its authors want clearly to attract attention to the conclusions of their research that more black children were being killed and injured on roads that were the everyday route of white commuters travelling from the exclusive suburb of Pointes to their workplaces in the city centre. Those drivers were accelerating their vehicles, in fear, anxiety or even rage, and putting the lives of black children in danger. Originally the map was part of a discussion paper, but it has now been reproduced in every paper and book about critical, counter and subversive mapping.

Inspired by the work of William Bunge, the counter-mappers of the twenty-first century are using similar cartographic tools in the making of official maps, including global positioning systems (GPS), geographic information systems (GIS) and crowdsourced data, to instigate a reflection on how these tools can produce very different results depending on the users' purposes. As Pickles pointed out:

> Bunge showed us [...] the truth claims of science must always be interrogated from the position from which you look, the shoes in which you walk and the maps you construct to guide your journeys.[53]

The New Zealand art critic Wystan Curnow proposes another interpretation for the emergence of such artistic counter-mapping practices, suggesting that art movements contested established concepts of urban cartography, much earlier than noted by scholars. Curnow points that the emergence of critical cartography in the 1980s was not necessarily pioneering, being preceded by experiments in art:

> It was while working on *Putting the Land on the Map* that I first read Wood and Fels's 1986 structuralist essay 'Design on signs: myth and meaning of maps' and was struck by the belatedness of its challenge to conventional views, for the contemporary artists I was interested in seemed to have already taken the point, whether or not they had read Saussure or Barthes.[54]

These artworks were bringing new paradigms to experience the city in its subjective, fluid and unfinished nature. Until the nineteenth century

[53] Pickles, John (2004). A history of spaces: cartographic reason, mapping and the geo-coded world. London: Routledge, p. xv.

[54] Curnow, Wystan. Mapping and the expanded field of contemporary arts. In Cosgrove, Denis (ed.) (1999). *Mappings*. London: Reaktion, p. 262.

cartography followed the tradition in landscape painting, representing the world on a flat surface and as seen from the close perspective of the observant in front (profiles) or above (panoramas) the subject depicted. The rupture with academicism and realism initiated by the impressionists also had an impact on the way maps were incorporated in art, starting with the dadaists in the 1920s inspired by the works of Baudelaire, Benjamin and Aragon, inciting the 'flânerie' or the bohemian wanderings in cities, as discussed in the previous chapter of this book. Developments in science also contributed to shake old beliefs and visions of the world, and the idea that its representation is an unbiased, objective, outside truth to be explored was in question.

Counter or subversive mapmaking can be found in participatory art forms that engage the public into the creative process. In the early 1970s artist Susan Hiller created a series of projects based on 'group investigations', using maps or mapping practices as prompts to perform artworks. In her 1972 piece *Dream Mapping*, Hiller proposed to work around the theme of borders where participants would dream and map in a specific location in England. The idea was to create a situation where known and unknown worlds would come together, relating spatial borders to the borders of consciousness, inspired by the very early forms of mapping that are still observed today in indigenous populations around the world, a well-known example being the Australian Aboriginal song lines that demarcate the paths and landmarks of ancestral generations in the Dreamtime. People would sleep outdoors in order to incubate dreams in a field with an abundance of 'fairy rings', commonly found mushroom circles in the countryside. This connects with other traditions in British folklore, as when falling asleep in a fairy ring one would be carried away to a fairy land, which in turn is a metaphor to be transported to an alternate state of consciousness. In another contemporary work, entitled *Street Ceremonies* (1973), Hiller invited the public to take part in a mapping exercise around central London.

Experimentations with public/performative mapping were also associated with political activism, as represented in the work of Trevor Paglen, an American artist and writer with a PhD in geography. His multidisciplinary practice, going back to the early 2000s, spans photography, sculpture, film making and investigative journalism, charged with a strong political content, focusing on surveillance and intelligence systems. One of his best-known works is *limit telephotography*, which uses astronomical photography technology to expose secret military bases in the USA that otherwise are 'invisible' to the public. It is representative of Paglen's questioning about the limits of democracy and participation, in a world where the politics of seeing dictates the extent of the knowable. Paglen is also regarded as having coined the term 'experimental geography' in 2002, to describe contemporary practices bringing together artmaking and critical geography, his two main fields of activity. In Paglen's words:

> Experimental geography means practices that take on the production of space in a self-reflexive way, practices that recognize that cultural production and the production of space cannot be separated from each another, and that cultural and intellectual production is a spatial practice. Moreover, experimental geography means not only seeing the production of space as an ontological condition, but actively experimenting with the production of space as an integral part of one's own practice.[55]

His definition is firmly based on Lefebvre's concept of production of space and brings not much novelty but, to its credit, anticipates the type of work that would characterise the 'map art' production of the early twenty-first century, as discussed by Denis Wood and others,[56] even if he does not make a direct reference to maps or mapping. 'Experimental' is also a useful word when referring to practices that are on the margins of representation and performance, encompassing the mix of art, mapping and activism. *Experimental geography* was turned into a book and corresponding travelling art exhibition (2008–2010), produced by Independent Curators International (iCI). Around the same time artist Lize Mogel (who had participated in the EG book/exhibition) and writer Alexis Bhagat edited and published *An Atlas of Radical Cartography*,[57] a collection of ten loose folded maps (including *CIA Rendition Flights 2001–2006*, by Trevor Paglen and John Emerson) and accompanying book with ten essays. Mogel and Bhagat define radical cartography as a practice that 'subverts conventional notions in order to actively promote social change', and the maps/essays were selected based on pieces that emphasise 'radical inquiry and activist engagement'.[58] In fact, the maps are all concerned with global contemporary issues, such as environmental crises, urban poverty, immigration and border conflicts.

[55] Thompson, Nato et al (2008). Experimental Geography: radical approaches to landscape, cartography and urbanism. New York: Enfield: Melville House, p. 31.

[56] See *Cartographic Perspectives*, number 53 (2006). Special Issue: Art and Mapping.

[57] Mogel, Lize and Bhagat, Alexis (2007). Op. cit.

[58] Mogel, Lize and Bhagat, Alexis (2007). Op. cit., pp. 6–7.

Trevor Paglen's radical geography points to the contradiction in the scientific claims of cartography to make all territory 'mappable' to the highest precision and accuracy, excluding areas for strategic reasons (not to reveal sensitive areas, not to reveal power, covert activities of the army and paramilitary) by exactly making images fuzzy and imprecise. Not only secrecy and political games are responsible for this but also economic reasons. It is still expensive to map with accuracy and areas are left behind, which justify the existence of initiatives such as Missing Maps.[59] Therefore, mapping with precision and accuracy parts of the globe that have not received the attention of institutional or corporate cartography is, in itself, an act of subversion, a type of counter-mapping that goes against the logics of political dominance and profit-oriented markets. In *An Atlas of Radical Cartography* architect, scholar and urban activist Jai Sen describes his work along with the Unnayan group on the settlements of people living in the margins of Kolkata, India, in a cartographic experiment in the 1980s, as 'making visible through mapping those who are traditionally made invisible by mapping'.[60]

As maps and mapmaking became more popular and widespread in the 1980s and 1990s, featuring across different media, in the art world we observe a growing interest in the field, with art curators and institutions giving 'map artists' more visibility and opportunities for exhibition.[61] New digital technologies developed from the 1970s onwards by artists and academics likewise converged in the exploration of the map as an aesthetic and informative medium to questioning, using technology for their purposes of experimenting with creativity. The novelty of these art, research and technological practices founded in the mid-twentieth century is the engagement 'from below' with the urban space, against the traditional detached view that homogenises perceptions. The experience of walking and everything that it brings with it; chance encounters, exposure to the elements, interruptions, feelings of boredom, excitement or pure tiredness can be incorporated into the mapping. In this way the map is never the

[59] *Missing Maps* is a humanitarian project that aims to map the most vulnerable places in the world subject to natural disasters, conflicts and disease epidemics, in order to help international and local organisations to respond to crises more effectively. It was founded in 2014 by the American Red Cross, British Red Cross, Humanitarian OpenStreetMap Team and Médecins Sans Frontières. For further information see https://www.missingmaps.org/.

[60] Sen, Jai (2007). Other worlds, other maps: mapping the unintended city. Mogel, Lize and Bhagat, Alexis (2007). Op. cit., pp. 13–25.

[61] Wood Denis, (2006) Map art. 5–14.

same. Towards the end of the 1990s and after, with a string of exhibitions, publications and a multiplicity of practices, the art map genre is established. Artists not only were incorporating maps in their artworks but also were putting mapmaking at the core of their practices. What was an occasional production turned into a constant theme, with the emergence of a generation of artists using cartography as their main medium and method.

In an article published in 2006, Denis Wood compiled a database of 218 'map artists', selecting individuals and collectives (active at that time) with significant production.[62] The sources of information were art exhibitions catalogues and books. As the author admits, it is an extensive list but far from being complete; nevertheless, it attests to the spread of the art map concept. A few years later, in 2009, artist and researcher Ruth Watson produced a paper analysing the growing interest of the art world in cartography.[63] She emphasises the 'explosion of maps in art' from the 1980s onwards, evidenced by her own list of 24 exhibitions of contemporary art, from 1997 to 2009. Reinforcing the argument of a surge in art maps, only 5 exhibitions are listed between 1977 and 1989, while the other 19 are between 1992 and 2009. Watson also notes that the list is not exhaustive—and might carry the same bias of selecting works from the English-speaking world, as Wood observed—but it demonstrates the prevalence of mapping in contemporary art.

Probably the most exemplary evidence of this new phenomenon is *The Map as Art: Contemporary Artists Explore Cartography*, published in 2009 by Katherine Harmon and cited in Watson's paper. The book presents a collection of 360 maps produced by around 180 artists and artists' collectives.[64] It is an assertion for the increasing widespread of artistic practices entwined with cartography along the second half of the twentieth century, as more and more artworks are produced towards the present days. The book includes five essays by art historian Gayle Clemans, who points out to remarkable characteristics, such as political instances and contestation, personal and emotional narratives, imaginary territories, cri-

[62] Wood, Denis (2006). Catalogue of map artists. *Cartographic Perspectives*, 53, pp. 61–67.

[63] Watson, Ruth (2009). Mapping and contemporary art. *The Cartographic Journal*, 46(4), pp. 293–307.

[64] Including three artists interviewed for this book: Christian Nold, Matthew Picton and Stephen Walter. Harmon, Katherine (2009). *The map as art: contemporary artists explore cartography*. New York, Princeton Architectural Press.

tiques of the digital age, all combined with acute creative visual appeal. Curators and institutions were inclined to exhibit these artworks not only because of an increase in production but also because they were coming in more varied, provocative and intriguing forms, introducing new materials, processes and themes. As Ruth Watson goes through the art exhibition's catalogues, she builds her rationale for what she calls an epistemological shift: from 'the map' towards 'mapping'.

In 2013 artist and scholar Karen O'Rourke published *Walking and Mapping: Artists as Cartographers*,[65] a book that documents a series of mapping projects developed by contemporary artists, many of them with the author's participation. As in Carieri's *Walkscapes*, published ten years earlier, O'Rourke credits the situationists as the precursors of the performative mapping, delved into the traditions of dérives and psychogeography. However, her book offers a further exploration on the contemporary experience of 'pedestrian-cartographer', describing how artists create and document their artworks, performing them individually or collectively. The concept of 'performance' is central to O'Rourke examination of such mapping practices, and she evidences the many elements derived from the performing arts contained in the artists' works. The use of technology is another characteristic, as many artists collaborate with information scientists and incorporate GPS (global positioning system) devices into their mappings to create 'datascapes', which is also used as a form of documentation, alongside photo and video cameras.

In the next chapter this age of 'mapping' in the twenty-first century will be examined in more detail through my conversations with artists about their practices, their motivations to incorporate cartographic techniques into their artworks and what meanings they infer from mapping cities. I also discuss some examples from visual artists focused on the 'urban pattern' and how they offer alternative narratives of cities while deconstructing the urban map.

[65] O'Rourke, Karen (2013). *Walking and mapping: artists as cartographers*. The MIT Press.

CHAPTER 5

Contemporary Artists Mapping Cities

The convergence of visual arts and cartography became more apparent at the turn of the twenty-first century, when 'art maps' appear as a label in the art world. As mentioned in the previous chapter, an increasing number of exhibitions and publications are testament of this phenomenon, and within this new scene, the number of artists engaged in mapping cities is considerably higher when compared to the sparse examples of preceding decades.

This growing interest might be firstly related to a revival of the situationists and their concept of psychogeography in the 1990s. In 1989 a major exhibition on the Situationist International was held at the Centre Georges Pompidou in Paris, before travelling to the Institute of Contemporary Arts (ICA) in London and from there to the Institute of Contemporary Art in Boston. The event coincided with the reprint of the Situationist International magazines, and the publication of catalogues and critical essays helped to spread once again the ideas of Guy Debord and his friends to a new generation of artists.[1] In the UK situationists and psychogeography had a particular comeback after that, starting with the

[1] In the case of the ICA in London the curators produced a catalogue-essay including the reproduction of selected SI documents from 1957 and 1962, accompanied by excerpts of British publications between 1966 and 1988, tracing the impact of the SI in British culture. Blazwick, Iwona and Institute of Contemporary Arts (1989). *An endless adventure. an endless passion. an endless banquet: a situationist scrapbook*. London: Institute of Contemporary Arts.

G. Lanci, *Art Maps and Cities*,
https://doi.org/10.1007/978-3-031-13306-0_5

reinstatement of the London Psychogeographical Institute in 1992 and the launch of its associated academic journal *Transgressions: A Journal of Urban Exploration*, in 1995. Despite being short-lived (both institute and journal came to an end in 2000), they marked the emergence of a renewed situationist movement, with the creation of art-activist collectives and walking groups.[2] This rediscovery of the Situationist International brought into evidence the performative character of urban mapping and the need to connect with space through an embodied practice.

Secondly, the imminence of the new millennium represented an opportunity to revisit the condition of the postmodern metropolis, sparking scholarly and artistic enterprises alike. The spatial vastness and social complexity of mega cities challenged the fundamental problem of redefining what a city is, and how to deal with the new reality of interconnected urban spaces, in the physical and virtual senses. Edward Soja, in the preface of his book *Postmetropolis*, embraces this challenge with a hint of enthusiasm:

> As we enter the new millennium the field of urban studies has never been so robust, so expansive in the number of subject areas and scholarly disciplines [...] so permeated by new ideas and approaches [...] It may indeed be both the best of times and the worst of times to be studying cities, for while there is so much that is new and challenging to respond to, there is much less agreement than ever before as to how best to make sense, practically and theoretically, of the new urban worlds being created.[3]

Envisioning what kinds of urban landscapes would emerge in the new century, artists and curators were prompted to take the city as a central theme to develop artwork and exhibitions. In 1997 curators Hou Hanru and Hans Ulrich Obrist organised an exhibition on Asian art focusing on the continent's frantic growth and modernisation. Combining works of artists and architects, *Cities on the Move* displayed a profusion of paintings, sculptures, videos, performances and installations, in and out of galleries' spaces, that attempted to mirror the intense urban development characteristic of Asian city-regions. The exhibition was equally set up to be on the

[2] For a comprehensive history of these initiatives, see Richardson, Tina (ed) (2015). *Walking inside out: contemporary British psychogeography*. London: Rowman & Littlefield International.

[3] Soja, Edward W. (2000). Postmetropolis: critical studies of cities and regions. Oxford: Blackwell.

move for two years: opening in Vienna in November 1997, it travelled from there to Bordeaux, New York, Copenhagen, London and Helsinki, generating new projects and collaborations in each location, before arriving in Bangkok as a 'city-intervention project' in November 1999.[4] The new Tate Modern in London, inaugurated in 2000, held its first major exhibition in spring 2001 entitled *Century City: Art and Culture in the Modern Metropolis*. As a celebration of the arts and ideas of the twentieth century, it was organised around nine cities, selected as 'crucibles for innovation', where the avant-gardes of arts and culture flourished during certain periods of the twentieth century: Vienna (1908–1918), Paris (1905–1915), Moscow (1916–1930), Rio de Janeiro (1950–1964), Lagos (1955–1970), Tokyo (1967–1973), New York (1969–1974), London (1990–2001) and Mumbai (1992–2001). The exhibition's companion catalogue, as voluminous as the art show was (it included 600 artworks and involved 12 curators), included essays from three eminent scholars on questions about western and non-western urban cultures, loft-living in megacities and the modern urban dweller. [5]

Finally, as the noughties advanced, the speculative and critical use of digital technologies became more apparent in the work of artists engaged in mapping practices. One reason for that is the availability and improvement of digital geographical tools like GPS and GIS, made more user friendly and intuitive to operate, but there was also an emergent curiosity among artists in questioning and investigating what institutions, governmental bodies and corporations were doing with such tools, and how they impacted people's lives. The interest in new mapping technologies started already in the 1990s and reflected a collaboration between the arts community, increasingly turning to the science of spatial data and the technology of GIS and web mapping, and geographers engaged with creative research through exploring art. Examples of the relationship between geography and art include the examination of 'creative geographies' in

[4] Swinson, James (1999). Architects dream of nightmare cities. *Third Text*, 13:48, pp. 101–106. See also Altshuler, Bruce (2013). *Biennials and beyond: exhibitions that made art history, 1962–2002*. London: Phaidon Press, pp. 342–354.

[5] Blazwick, Iwona (2001). *Century city: art and culture in the modern metropolis*. London: Tate Gallery. Ziauddin Sardar, professor of postcolonial studies at the City University; Sharon Zukin, professor of Sociology at Brooklyn College; and Joachim Schlör, professor of Jewish studies at University of Southampton, authored the essays.

scholarly research[6] and the inclusion of 'experimental geography' in art practices.

This trend has relations with the emphasis on inter- and transdisciplinary research. Back in 2000 Denis Cosgrove and Luciana Martins pointed out the dissolution of hard boundaries between artistic and scholarly practices, in humanities and science, saying that 'creative artists increasingly draw upon, and contribute to, the practice of geography, while geographers themselves are adopting more creatively imaginative ways of making and communicating knowledge'.[7] Kathrin Busch, in 2009, identified 'a trend in contemporary art to focus on the production of knowledge', which is resulting in art productions 'characterized by an interdisciplinary procedural method'.[8] However, the spectrum of what could be named 'artistic research' is quite broad, she admits, including a variety of forms and aims.

The discussion that follows does not intend to build an argument on whether these events—the revival of psychogeography, the twenty-first-century city as privileged subject in both art and research and the consequences of the development of web-based mobile cartography—have a direct influence on artistic practices, in the sense that they have shaped a specific style or trend in contemporary art mapping. What can be argued, though, is that they constituted an enabling environment for artistic practices towards mapping cities, and they formed a common cultural and intellectual background shared by the artists I have interviewed and others whose works are commented in this chapter. That being said, it can be observed that for each artist one or more of these events weigh differently in their practices, being more influential or totally dismissed, and their impact must be understood against the artists' personal trajectories. What is more significant in the practices of contemporary art is that most of the artists pay great attention to the mapping process, beyond being

[6] There is a good range of academic papers and books around the theme; to cite a few, Dear, Michael (ed) (2011). *GeoHumanities: Art, history, text at the edge of place*. London: Routledge; Hawkins, Harriet (2013). *For Creative Geographies: Geography, Visual Arts and the Making of Worlds*. New York: Routledge; Marston, Sallie and De Leeuw, Sarah (2013). Creativity and geography: Towards a politicized intervention. *The Geographical Review* 103: iii–xxvi.

[7] Cosgrove, D. and Martins, L. (2000). 'Millennial geographics'. *Annals of the American Association of Geographers*. 90 (1) pp. 97–113.

[8] Busch, Kathrin (2009). Artistic Research and the Poetics of Knowledge. *Art & Research*, vol 2, n 2.

exclusively concentrated on developing new forms of art maps, and how to critique standard concepts of place and space through their artwork. Receiving the legacy of the twentieth century in artistic experimentation, artists now have the opportunity to play with cartography in a trans/interdisciplinary research environment, where embodied experiences can be associated with the construction of spatial knowledge.

Together with the theoretical speculation of the first four chapters, this one provides the general context and framework for the analysis of the case study of Liverpool that follows. In the first section, I will present my in-depth conversations with four British artists on the search to find out what motivates them to produce their works, which methods they use and how they place themselves on the intersections between mapping and artistic endeavours. They are from a generation that engaged with maps and mapping after 2000 and represent a varied group in terms of artistic practices and techniques, using sculpture, collage, drawing, performance and participatory mapping. In the same way they are not restricted to exclusively mapping urban environments (as none of the artists I previously discussed are) or even to producing maps as their main practice, but in common they indicate through their work similar interests in exploring the possibilities of mapping, challenging the aesthetic qualities of maps and their use as vehicles of geographical 'truth', questioning systems of visual representation, discussing mapping practices in everyday life and the politics of space, and reflecting upon the role of art and mapping practices in the urban context. In the second section, I make some brief comments about five contemporary artworks that explore the urban pattern, which has often been mapped throughout history as systems of street networks, as the city is viewed from above. These artworks are presented here as examples of creative ways to discuss the previously defined 'viewing from above' perspective and to further question what spatial representation means. They are characteristically produced in series, expressing a fascination, if not obsession, of the artist with mapping methods and the endless variations they allow, which suggests an imperative immersion of the artist into the subject: some map series are produced over many years. Although using different techniques and approaches to mapping, the artists bring forth deep sensory connections with cities that are enabled through the embodied mapping exercise while keeping away from a disembodied objective practice of cartographic representation.

Four Conversations with Artists

Creative engagement with mapping, beyond the constraints of precise geographical descriptions, offer artists ways to extract other meanings from cartographic techniques, incorporate interpretations of places that can combine narratives and myths from different times in the history of human civilisations or bring together distant places in the planet to the same location in a unique map frame. Decoding, displacing, dematerialising: from the artistic perspective, the possibilities to explore visual interpretations of the territory are endless. In my encounters with artists I could observe, register and learn how and why they convey such meanings, discussing their methods and techniques through their artworks.

In the following pages I introduce the work of four artists, starting with brief descriptions of their practices and production, to then further discuss the main themes that emerged from our conversations. These were carried out in 2017 through face-to-face interviews in the artists' studios (Layla Curtis and Ruth Levene) and residence (Christian Nold), with the exception of Matthew Picton, who I talked with via a video conference. They took the format of informal talks that on average lasted two hours, which were fully recorded and later analysed to extract material for this book. As mentioned previously these were semi-structured interviews, with some key questions around their careers and how they started incorporating mapping into their practices, their knowledge and interest in cartography in general, which methods and techniques they employ and what meanings or messages their work intend to communicate. The questions were rather used as guidelines, and I deliberately let the artists speak for themselves and conduct the flow of our conversation. This resulted in dedicating more time to explore their artworks in detail, discussing them in descriptive and interpretative ways, examining the original pieces stored in their studios, as was the case of Layla Curtis and Ruth Levene; printed maps and books, in the case of Christian Nold; and online digital websites and archives, which was useful in all interviews and particularly in the case of Matthew Picton, as we went together through his works compiled in his website.

The artists interviewed have works identified as 'visual art', but they also experiment with sound, performance and installations, using diverse techniques and media. None of the artists interviewed refer to themselves as 'map artists' or try to label their work within rigid categories, but they share a similar approach to the incorporation of mapping into their practices as an opportunity to reflect upon art processes, and a way of thinking and theorising about place, space, history and landscape, rather than focusing on the construction of maps as aesthetic objects devoid of meanings. Analysing the recorded talks afterwards I could identify three predominant themes during the interviews, which I organised in subsections: the boundaries between art and science, mapping as embodied knowledge and the power of maps in transforming perceptions of the urban.

Layla Curtis

Layla Curtis studied at Edinburgh College of Art and at Chelsea College of Art and, since 1999, has been based in London. Her work often explores notions of borders and boundaries, questioning the geographical 'truths' of representation and cartographic paradigms. She works largely with drawing and collage to compose her maps, but her practice also includes video, photography and digital mixed media.

Curtis makes collages out of existing maps with an interest in examining them as constructed artefacts and at the same time examining the presumptions that they encourage. Just after her graduation in 1998 she took up residency at the Akiyoshidai International Arts Village, Japan, where she developed her collage technique, involving painstaking work cutting and reassembling Japanese road maps. During the same period, she produced *United Kingdom* (1999), which consists of two panels: the larger one has the outline of England, while the smaller one has the outline of Scotland, but it is placed on the upper left corner, as an inset, occupying the location of the Irish island. On first glance these are apparently reliable cartographic representations, but on looking closer, one is faced with disorientation: in the English landmass are cities and towns from Scotland, Wales and Northern Ireland, while the Scottish island is made of English, Welsh and Northern Irish maps. Within each panel urban areas are also exchanged, with coastal cities occupying the hinterland and vice-versa.

This early experiment with collaged maps is representative of Curtis' mapping work, deceiving the eye and playing with visual cognition and, ultimately, geopolitical knowledge. Her invented cartography is much related to ways maps convey information through visual language and codes. In *NewcastleGateshead* (2005) (Fig. 5.2) a similar strategy is employed, this time recreating the northeast of England from international atlases. Urban areas were cut out from 'Newcastle' and 'Gateshead' cities around the world, as well as homonymous towns, rivers and regions in other countries. In 2013 Curtis constructed a ten-part collage map for the Thames Festival using topographical maps from around the world with namesakes connected to the history of London's main river (Fig. 5.1). For the duration of the ten-day Thames Festival, the work was available as a free digital download to those visiting the Thames Festival website. It was also printed on large banners exhibited on the riverside walkways in front of Tate Modern and by Tower Bridge.

Around 2001 Curtis started to work with hand-drawn maps, later producing a series of urban maps called ***Index Drawings*** that includes London, Paris, New York, Glasgow and Edinburgh (Fig. 5.3). The series is based on street indexes, which in the UK were made available on maps in the mid-1800s and later became the companion lists in the backs of A-to-Z map books of the early 1930s. In Curtis' drawings the layout of cities is delineated by street names; all other information and graphic elements are omitted. Again, viewed from a distance, the maps bear a familiar shape with traditional guide maps, but on closer inspection the stratagem is revealed.

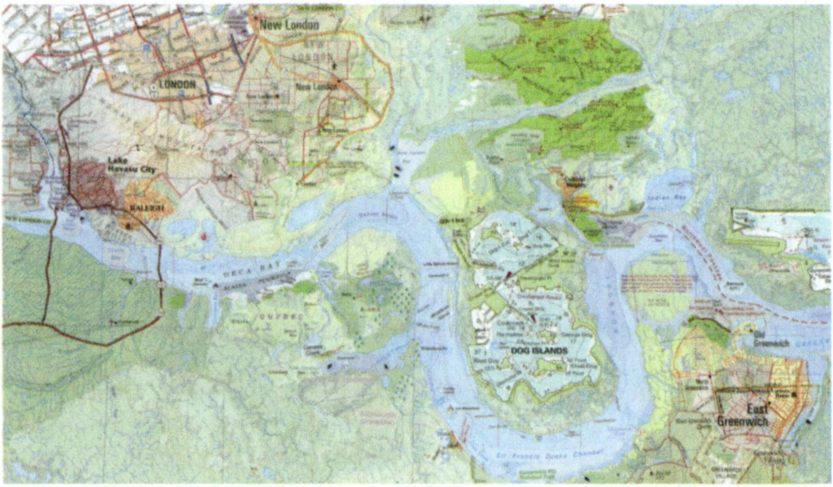

Fig. 5.1 *The Thames* (from London Bridge, Arizona, to Sheerness, Canada) (detail). Layla Curtis, 2013

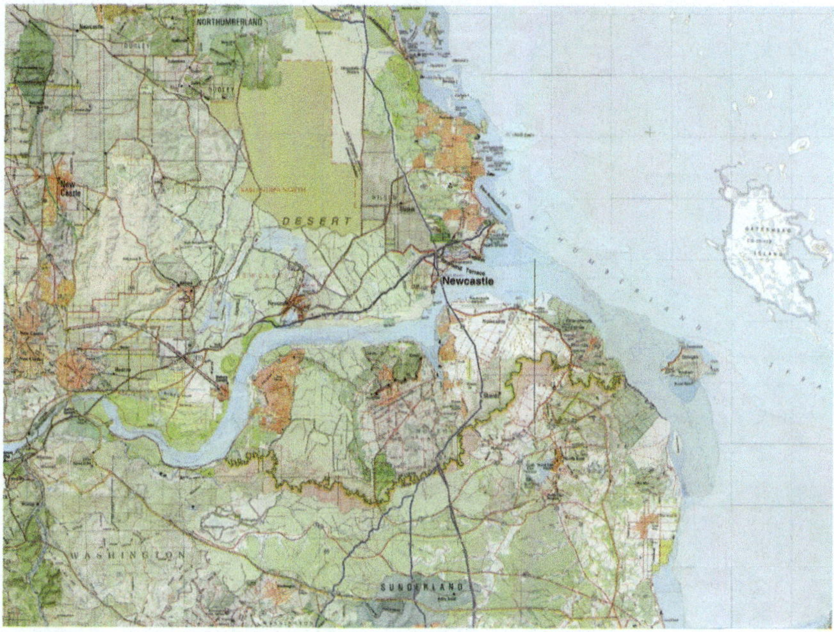

Fig. 5.2 *NewcastleGateshead* (detail). Layla Curtis, 2005

Ruth Levene

Ruth Levene studied at Hull School of Art and Design, graduated in 1998, and is currently based in Sheffield. Her practice involves research led participatory and socially engaged work, and it covers a range of media, including video, digital drawings, performance and installations. Maps and mapping were present in her practice from early on, when back in 1999 she started to collect hand-drawn maps given to her from people showing directions or telling stories and indicating places. But it was not until 2008 that the collection came to be exhibited as *Hand Drawn Map Collection* (Fig. 5.5).

In that same year Levene took a residency in the Yorkshire Sculpture Park (YSP), conducting a project together with a surveyor from the Ordnance Survey. The result is a series of maps entitled *As Far as the Eye Can See* (Fig. 5.4). The maps were manually plotted from pre-selected viewpoints spread around the park area, where the artist imposes her view, metaphorically and technically. Her sight is determined by her physical position, her eye's height in relation to the terrain and the weather conditions over the several days that were required to complete the project.

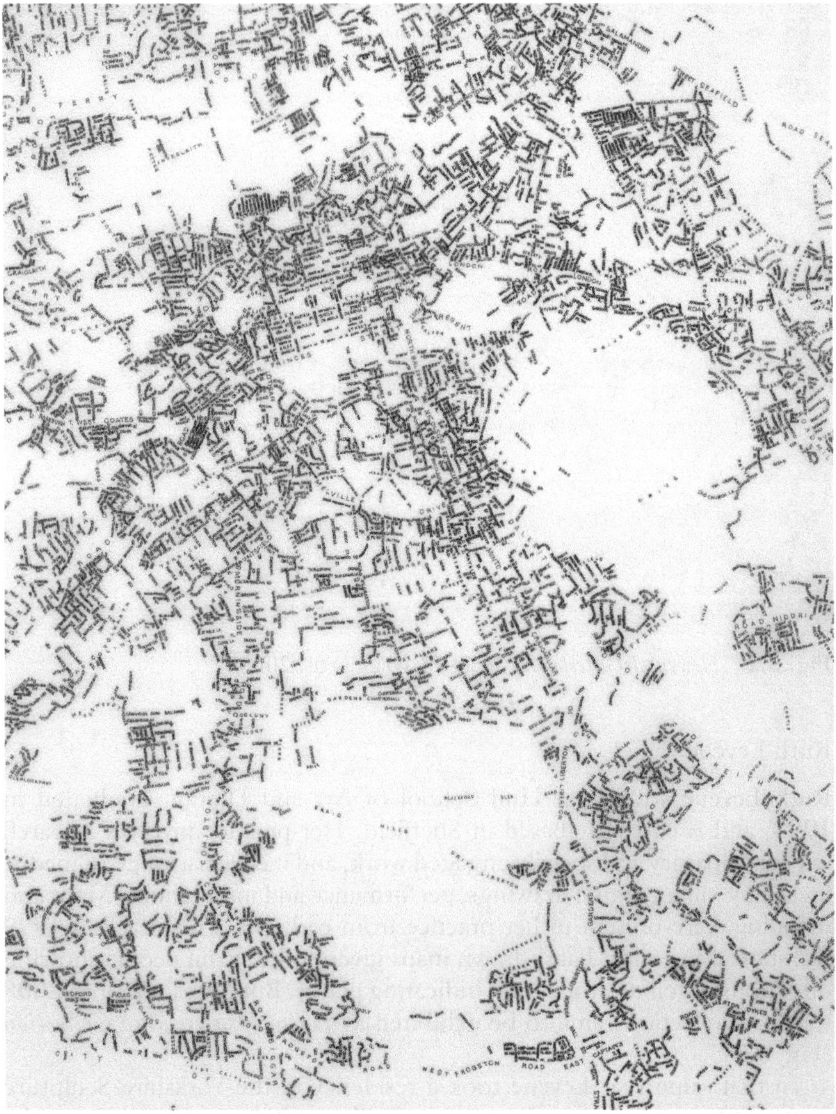

Fig. 5.3 Edinburgh Index Drawing (detail). Layla Curtis, 2012

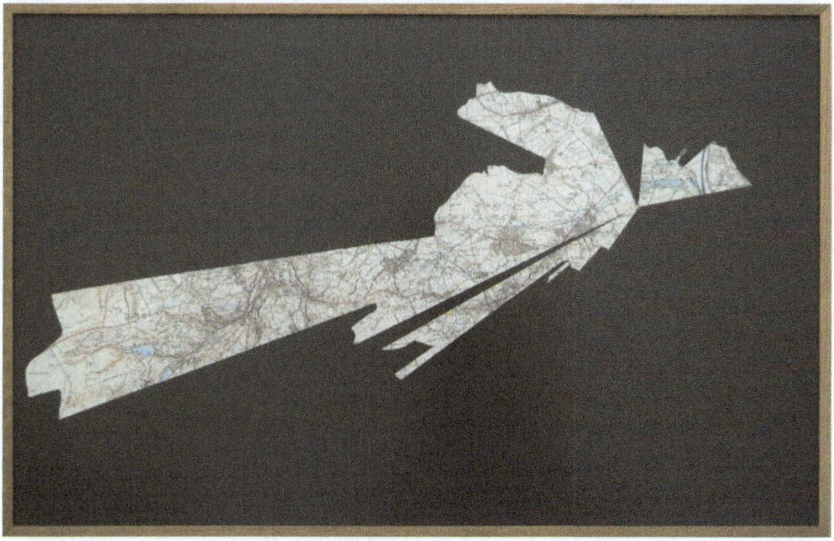

Fig. 5.4 *As Far As The Eye Can See* (frame 1/6). Ruth Levene, 2008

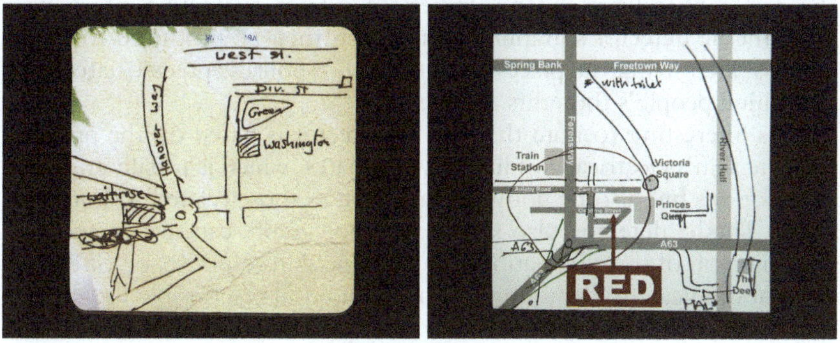

Fig. 5.5 *Map Collection.* Ruth Levene, 2008

These conditions enforced the cartographic rules to build maps that correspond to her own visibility of the field, at a certain time.

In documenting urban spaces Levene experimented with different modes of mapping, such as walking the streets of Brooklyn, New York, with a voice recorder and a notebook, registering her route along the way and later rendering this over a commercial street map of the city. More recently she has been engaged with environmental debates where the urban space is approached in connection with the wider landscape and the natural resources that eventually shape its form. An example of this is *Hidden Waters* (2015), a series of digital drawings and etchings mapping the 'invisible' urban water supply and sewage infrastructures, alongside river systems of Britain.

Christian Nold

Christian Nold is an artist and researcher based in London who is involved in developing participatory projects for communal representation. Nold attended the Royal College of Art, graduating in 2004, since when he has been working in collaborative projects worldwide, especially conducting his most well-known initiative, ***Bio Mapping***, a method for visualising people's perception of urban environments.

The first bio mapping experiment, formerly called 'emotion maps', was undertaken in the neighbourhood of Greenwich, London, in 2005. His method consists of inviting people to go for a walk wearing an electronic device that combines a lie detector sensor and a global positioning system (GPS) that records the wearer's geographical location. The data collected indicate peaks of emotional arousal, which is discussed collectively among the group of walkers and researchers to extract qualitative information from the lie detector's graphics. A map can then be produced from this process, identifying the places of emotional responses and associated texts describing people's thoughts and feelings.

It is interesting to note that the lie detector is based on the principle that the skin's electrical conductivity fluctuates, which is usually triggered by emotions but also by body changes such as those related to hormonal levels. The lie detector, also known as a polygraph, can record and produce a graph of this fluctuation (the galvanic skin response, or GSR), but it cannot indicate the emotion or body condition responsible for changes in the records, and for this reason an interpretation of the data is needed. Therefore, Nold's artwork relied on giving meaning to the data through the discussion with participants, facilitating the interactions that lead to a more in-depth perception and analysis of urban spaces. In this sense, the polygraph is a fulcrum for enhancing participatory experience and ultimately 'a participatory urbanism'.[9] Since 2005 the project has been conducted in various cities around the world, involving thousands of participants in workshops (Fig. 5.6).

Matthew Picton

Matthew Picton is a British artist based in the USA who explores history, narrative and place in three-dimensional form by constructing elaborate paper sculptures. His work is mostly inspired by documented historical episodes of political turmoil, societal change and cultural significance in cities, and by fictional or mythological characters, extracted from classic and modern literary texts, poetry and films, that evoke specific urban spaces and stories.

Picton was born in London in 1960 and obtained a degree in politics and history from the London School of Economics. His interest in art began just after finishing his studies, when he started painting and pursuing a career as a visual artist. Moving to the USA in the early 1990s, he settled in Ashland, Oregon, where he still lives and works. He has exhibited widely in the USA and Europe, and his work is included in the collections of American art museums (the Portland Art Museum and the De Young Museum, in San Francisco), of museums and institutions in the UK (the Herbert Museum of Art, in Coventry, and the Fidelity Bank, in London) and the Stadt Museum in Dresden, Germany. He has also undertaken commissions for hotels, hospitals, banks, museums and private clients.

Picton started producing city maps in 2006, initially working with a variety of materials, including paint, resin and enamel. After this first phase he made more than 20 maps of cities in Europe, the USA, Middle East and Asia, which evolved from flat surfaces of street networks to a more three-dimensional representation, arranging street networks over background layer of rivers and other physical features. Eventually Picton found in paper his preferred medium and started to produce his unique sculpted city maps around 2008, applying techniques like printmaking, painting and collage.

[9] The project aims to transform the lie detector into a different way of staging physiological arousal that facilitates a new kind of urban body articulation that becomes a participatory urbanism. Nold, Christian (2018). Bio Mapping: How can we use emotion to articulate cities? *Living Maps Review*, 4. Available at http://livingmaps.review/journal/index.php/LMR/article/view/103.

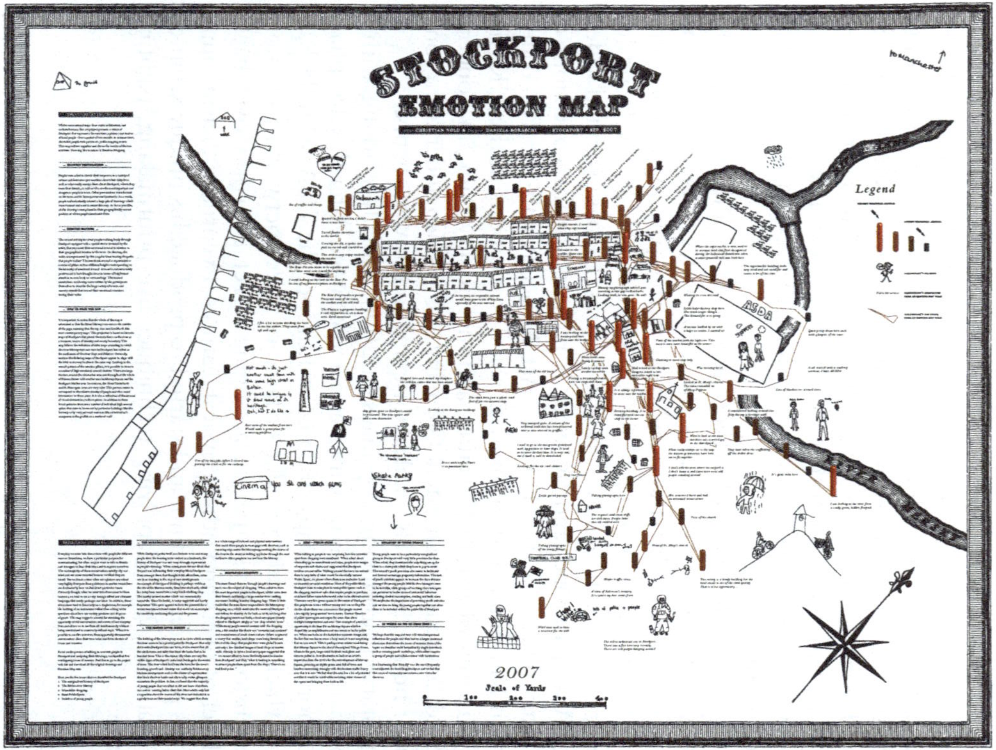

Fig. 5.6 *Stockport Emotion Map*. Christian Nold, 2007. The columns mark the points of emotional arousal; taller columns indicate more intense arousals. The drawings and texts help to qualify the experiences of people on the ground, such as 'feeling a bit uncomfortable here' or 'I can smell and hear the market's atmosphere'

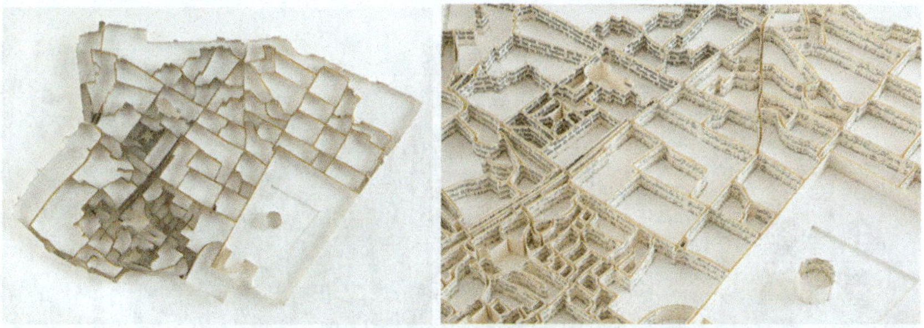

Fig. 5.7 *Jerusalem*. Map (left) and detail (right). Matthew Picton, 2011

There is an interesting play with information in Picton's work; some maps are deceptively simple but full of symbolism. The map *Jerusalem*, made in 2011 (Fig. 5.7), presents a blank orthogonality when viewed directly from above, but in an oblique viewing, it reveals the text extracts printed on the paper, taken from The New Testament, The Torah, The Armenian Bible and The Koran, placed accordingly to the geographical four religious quarters of the city. The texts produce differences in colour and shades in the map, but visually the borders cannot be clearly identified, as if the old city of Jerusalem persists as an entity despite its divisions. The map *Lower Manhattan* (2011) (Fig. 5.8) looks rather homogenous at first sight, but a closer inspection reveals that it is undoubtedly referencing the events of the September 11 attacks, created using newspapers published in the aftermath, the covers of the film "Towering Inferno" and book covers of the novel *The Plot Against America* by Philip Roth. Words like 'bastards!', 'terror', 'horror' and 'attack' are spread out through the sculpture, and on the site of the World Trade Center the map is charred black. The symbolic layers of explicit texts and the general greyish appearance of the map conjure the mixed atmosphere of shock, revolt and uncertainty that convulsed the country and the world due to the attacks.

Other maps explore fictional and fantastic aspects of urban spaces. An example is *Dublin June 16 1904* (2011), produced entirely from the novel *Ulysses*, by James Joyce, one of the most iconic narratives of urban landscape in modern Western literature. In the map *Los Angeles #2 Bladerunner* (2015), Picton constructed a completely different scenario, using as bases the poster of the film 'Blade Runner' and book covers of noir fiction set in the city or from authors related to the city. The sculpture also includes lines of text and quotes extracted from books about the history of Los

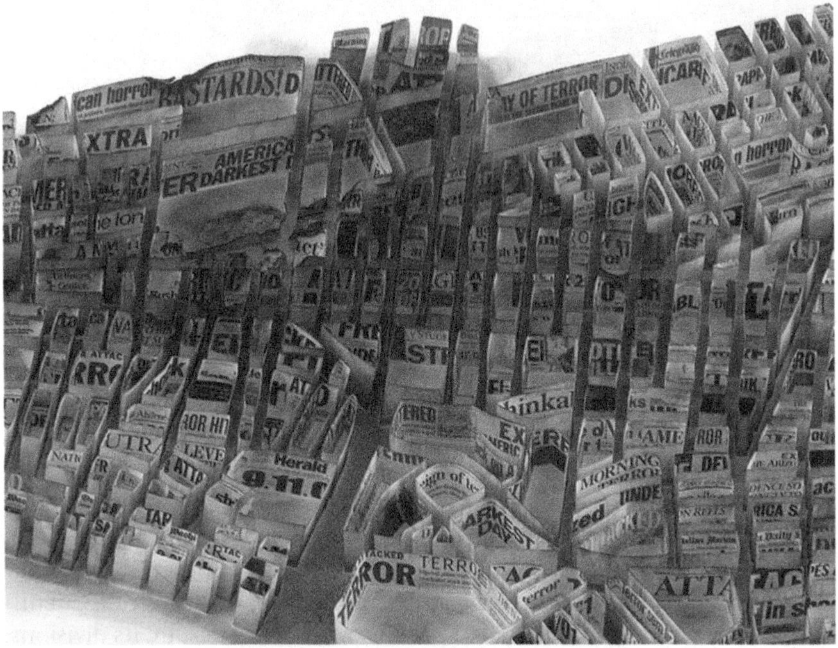

Fig. 5.8 *Lower Manhattan* (detail). Matthew Picton, 2011

Angeles and the now classic study *City of Quartz*, by American urban theorist Mike Davis.[10] Overlaying fiction and nonfiction, the work captures the type of 'schizophrenic' urban imaginary of Los Angeles, built upon the postmodern culture of late twentieth century.

Accuracy and Authority on Maps

Ruth Levene considers *As Far As The Eye Can See* the work that is most closely related to the concept of subverting traditional approaches to map and mapmaking. The project is based on a series of pre-selected points spread around the Yorkshire Sculpture Park (YSP) from where Levene and her collaborator, a surveyor from Ordnance Survey (OS), triangulate the artist's position on the ground and mark on the OS map the perimeter of the furthest horizon line that she could see, with the naked eye, in a

[10] Davis, Mike (1990). City of quartz: excavating the future in Los Angeles. London: Verso.

360-degree turn. The work took several days and therefore observation was subject to weather conditions, but this was indeed more than a literal observation of the landscape, as Levene emphasised:

> This is not a mapping product; this is an act of looking. My seeing skills and my looking skills over those two weeks were so exercised, they were so refined. I could spot things and know what I was looking for and seeing in the distance in a way that I have never looked before.

Apart from the tools for triangulation, the OS maps and pencils, no other equipment was used. The five resultant maps, precisely plotted to correspond to their coordinates in the original map, are an imposition of the artist's view—metaphorically and technically—over the territory, which was subject to weather conditions, the artist's sight and her position in the field.

The process of making *As Far As The Eye Can See* also uncovered diverse judgements in relation to art and science. The exchange between the artist and the OS surveyor brought to light that common perceptions of what is artistic and scientific in maps and mapping were too narrow. As they developed the work together on the field at the YSP, the surveyor became increasingly aware of how art could be more closely related to science and critical thinking than he supposed, while the artist realised that her own view on the accuracy of the OS maps is dependent on individual choices taken on the field, technical hindrances or economic decisions. As Levene describes the experience:

> That process at YSP was interesting for the both of us [...] I guess, naively, my perception of OS maps was based on their authority, their accuracy, their truth, and that kind of crumbled. His [the surveyor] perception of what art could be also changed: he thought that art was expressive and emotive, to discover that it could be more technical, rigorous and critical than he ever imagined. We went on this journey of crossing paths, where we gained an incredible respect for each other's practice. The [art / science] boundary just melted, it was blown away.

Christian Nold takes on the debate about art versus science vehemently: for him, making a clear distinction between what is supposed to be in the realm of art and what in the realm of science is 'very crude', suggesting that the focus of the discussion should be somewhere else:

> The question for me is not between art and science; it is between practices that have an actual meaningful relationship to place and relations, and practices that are totally disconnected and abstracted, relating to institutional models of power. I think that you certainly get some art that does that, and you get some science that does that.

Nold is talking about how he could position his own work (neither art nor science) but also how to interpret practices that are outside of 'readymade categories', as he named them, of the classic canons of art or science. This type of art-science debate is reminiscent of the difficulties in freeing art from its historical attachment to aesthetics that was entrenched in the nineteenth century, but it is also a persistent theme in the contemporary world because so much of what has been experienced in art is related to technologies that traditionally were in the domain of science. Denis Cosgrove already commented in 2005 that art and science shared epistemologies observed in pre-modern images that 'remained obscured [...] by claims for representational objectivity developed within modern science'.[11] This is better exemplified in some criticism that Christian Nold received, suggesting that his mappings are not very 'scientific' due to the lack of demographic control. Nold's reply is that this is a type of 'naïve criticism', made under a strict sociological view of his project:

> People who have used my data the first thing they do is try to divide between men and women [...] I think this is the real danger of sociology [...] these dangerous readymade classifications onto things in order try to understand and capture the world. It is a very particular top-down institutional framing of what publics are and how publics are constituted.

This 'invasive construction of the public', as Nold says, is a reflection of the separation between art and science that does not consider the ambivalences of technology that 'can enable any of these types of framings', according to Nold. The question remains in reductionist views of art being strictly concerned with the subjective approaches of reality and science focused on objective representations, as Cosgrove stated. This is more obviously apparent when dealing with cartographic accuracy. Accuracy though is often appropriated by artists from authoritative maps produced by institutional mapping agencies such as the Ordnance Survey, which

[11] Cosgrove, Denis (2005). Maps, Mapping, Modernity: Art and Cartography in the Twentieth Century. *Imago Mundi*, 57:1, 35–54.

indeed served as base maps for two of the Liverpool maps that will be discussed in the next chapter. However, the assumed accuracy in projection and scale borrowed from official sources is not the final objective but a point of departure to explore other aspects of the map.

Matthew Picton seeks to make his maps spatially and temporally accurate:

> I am a firm believer in artistic license and interpretation - but I would tell you that I do strive for accuracy, typically in the maps I have made. Well, dare I say it, they are very accurate probably. They are not completely accurate, the roads might be slightly bigger or slightly smaller, but they are true to the form, for sure. In terms of the inclusion of the subject matter I take pains to locate it in the right time period as well.

Christian Nold, by contrast, does not regard accuracy as essential in his work:

> I do not think people question the accuracy of the maps I make, because I certainly try to heighten the process of making them. I think it is hard to imagine that they are treated in the same way as a globe or something.

The difference here is not an observance to a standard of accuracy of scale and projection but the appliance of some aspect of accuracy that makes the mapping more meaningful. In Picton's maps, for instance, the 'form' he aims for will provide the background for other elements that are more significant in the map, as he makes the point that he is not interested in 'the exactitude of location' of historical sites and landmarks; in Nold's mapping practice, the accuracy is there for guidance of navigation and registration only. In both cases the use of the 'official' map releases the artist to explore and seek adventure in the urban form beyond formal designs.

This process of 'deconstruction' of accuracy does not suggest erasing geographical references. Implying that art maps dismiss accuracy or are purposefully inaccurate is falling into the trap of binaries, or a dialectics of opposites. The point is that accuracy is not a goal but a way to make the map resemble the urban pattern—it is the backbone of the mapping exercise that will make it possible the map to be recognised as a city map. There is the need for something that hangs on the map's visual language that indicates to the observer that the artwork is related to a city and not an amorphous landscape.

Cities in art maps have been explored through repetition and seriality, in reproducing particular aspects or urban landscapes, and more often by means of playing with techniques, media and established concepts of cartographic representation. 'Truth' is subject to different points of view, and 'reality' turns into an abstraction. As Ruth Levene observes, they are relative concepts that have long been manipulated in visual arts:

> You can go back to Picasso on this one: if I shut one eye or another, I just see a slightly different view. This 'truth' obsession, or the accuracy of the representation, is always ripe for fun.

Describing how she conceived the Thames map Layla Curtis provides an insight on her method to create an artwork that plays with the ambiguities of cartographic truth:

> [...] making sure that it is convincing is the biggest challenge; there is no point in just arranging them [*the cut out maps*] all in the shape of the Thames. The roads must connect, otherwise it is too fragmented and no longer believable as a map [...] at the same time, I like viewers to slowly understand how it is made, to know it is a hand-constructed collage. I kind of play with that balance between retaining a sense of the collage and creating a believable map.

Making choices to produce an urban map will invariably involve dealing with the selection of information that indicates that the artwork is a piece about a city. Usually, it is the street network design that gives the clue, but other elements can come into play, in the form of text and images that refer to the urban environment. For the artist there is a challenge in making the map 'believable', as Curtis points out, but there is also an opportunity to present multiple interpretations, to translate reality in ways that show 'truths' rather than 'the truth'.

The absence of information in Layla Curtis' *Index Drawings* is revealing of spectral traces of the city, and a similar effect is observed in the GPS maps of Jeremy Wood. It plays with the perception of the viewer, who instinctively fills the 'gaps' through the power of imagination: a large empty area can be a park, a football pitch, a woodland; wide and long, continuous lines can denote a river, or very straight and narrow can be a railway. The drawings invite a mental exercise, but certainly to be able to visualise these elements—the park, the river or the railway—one must be familiar with the representations of the urban landscape from above.

Maps like the *Index Drawings* are playing with the asymmetry between the map and the territory, the coded language of symbols and the coded language of spatial cognition. Because the street names are extracted directly from A-to-Z guides, and not from commercial maps, the *Index Drawings* expands the lexicon of the street guides in creating a visual script of the city. The shapes that are formed by the writing of the street names are the only indication of urbanity, as the writing itself follows the unvaried and monotonous character of the original street name lists, not disclosing any signs of hierarchy or place significance. Reduced to no cartographic information: no scale, no relief, no lines or polygons, no marks, no legend, no title, nothing that informs it to be a map, but we still recognise it as a map, a map of a city:

> All the usual graphic components of a map have been taken out, or […] have not been put in […] but these works still function as maps. That is what I am playing with here: just how much information do you need to be present on a map in order to be able to locate yourself or navigate successfully? […] The information that I have stripped away is also telling you something, and that is the magic of these drawings.

The use of Ordnance Survey maps, or the A-to-Z guides, does not acquiesce to the supremacy of institutional mappings, but rather it interrogates the authority they impose upon the territory, by acts of literal deconstruction (e.g. the cut-and-paste collages) and other nuanced "misuses". Ruth Levene, for instance, points out that in her art practice mapping is a means of investigation, a method of enquiring:

> It is a kind of tool, the process of mapping for me. There is also this curiosity about the actual concept of a map; there are many ways to explore, change, alter or possibly undermine, subvert the map, to question the authenticity or the authority of the map.

Playing and tricking the mind through visual alterations, by disrupting or ignoring the rules of cartography, exposes the limitations of geographical knowledge when subject to alterations and distortions; at the same time this playfulness enables the observer to think about the city as a space of constant flux rather than unchanged reality.

Embodied Knowledge and the Politics of Space

Other experiences in mapping conducted by artists give interesting insights into how mapping on terrain can reveal embodied knowledge of the landscape that otherwise would not be apparent. For the execution of her artwork *Tong Tana* [2012], the artist Layla Curtis travelled to Borneo to capture in video the journeys of a Penan hunter-gatherer man into the rainforest. Immersed in an unfamiliar environment for her and relying on being guided by the local community, Curtis described the jungle as 'a wall of vegetation and you hack your way through [...] it is really a three-dimensional experience.' Among the group there was one man, Denis, who had never been to that place in his life and did not know the landscape, who was able to draw six maps over the six days' journey as they moved on through the forest:

> I know how to draw a map, but I couldn't have drawn *that* map. I had assumed we had been crossing lots of different tributaries running into a larger river in a valley below us. However, when I asked Dennis (my Penan guide) to draw a map of our route for me he drew a river with many meanders, and I realised that for the whole day we had in fact been crossing the same winding river. Of course, this might have been more obvious to me if I had been paying attention to what direction the water was flowing.

His maps were 'incredibly accurate', according to Curtis, who compared them with satellite images and the route she had traced on GPS. His maps showed her that 'local knowledge on the ground' associated with an instinctive sense of navigation gave this man a unique mapping skill. As we were talking about the Borneo project, I raised the question 'how would that compare to an urban environment?', to which Curtis responded:

> [in the forest] I miss the details and clues in the flora and fauna as I am not attuned to them, and so heavily vegetated places very quickly have the tendency to look similar. However, as a city dweller, I can notice the tiniest of details, and these sub-conscious micro-observations help me navigate through the urban environment. It was soon obvious that I did not yet have the knowledge to transfer these skills to the forest. But I was with people who had lived there all their lives and could unsurprisingly instinctively navigate through parts of the forest that were new to them. That felt very exciting to witness.

Such observations reflect how the innate navigation skills that human beings have can be perfected over time by just living in the same environment for a long period, building up a catalogue of references that help to create such as a mental 'compass'. The ways in which that translates into a mapping language will depend on the ability to communicate spatial knowledge, which often is resourced in some already constructed mapping visual code, such as using lines to represent paths/streets, or polygons representing open areas/buildings and points representing junctions/landmarks.

The hand-drawn maps collected by Ruth Levene are an example of the inherent quality of mapping as a means to capture the instinctive skills of navigation and orientation. The artist's role is to construct a narrative, to evidence these qualities of intuitive mapping. When Levene was composing the collection for an exhibition, she observed that they were:

> [...] revealing of how people transfer information: some would be very acutely aware, very neat, and would choose carefully what information to put in and what to keep out.

Levene's artwork is reminiscent of an experiment devised by conceptual Dutch artist Stanley Brouwn in the early 1960s. Brouwn approached the public on the streets asking them to draw directions to a particular place; he would then produce a series of prints from the hand drawings stamped with the text 'This Way Brouwn'. Although his encounters with people would happen randomly, he took care to register every single one, and on those occasions when someone did not draw anything, Brouwn would include a blank page to the collection, with only the stamp on it.

Representation is, however, just one aspect of the spatially embodied knowledge. Christian Nold's mappings are based on the embodied experience, using sensing technologies that are literally attached to people's bodies. It explicitly interrogates how much of what can be recognised as spatial perception can be related or transferred to a map. Nold observes that what people perceive and respond to walking around in the city corresponds to the way people apprehend the complexity of the urban space:

> It is a mixture of things that are inside your body and things that are outside. People start to re-think where their body begins and ends, and where the environment begins and ends.

The map in the context of Nold's experiment is just a 'metaphor for making sense of the data', and it is the collective interpretation of the data that reveals the true mapping exercise. The artists are also aware of how mapping reflects into political stances in their work, as they acknowledge that in mapping cities it is difficult to avoid the theme of politics. As Matthew Picton observes:

> Any city has, inevitably, a political component [...] Of course you can look at a city and decide to focus on different elements that are not particularly political. But usually, and certainly it is when becomes more interesting, there is going to be a political component. That said, I am not always attempting to make a political statement when I am making a piece of art.

Other artists similarly acknowledge the role of politics in their practice; it might not be explicitly stated in their pieces, as Picton recognises, but it is clear for them that all cartography has been historically political, and this legacy is intrinsically connected with what they do. Ruth Levene, for instance, affirms that her work 'definitely' has a political aspect to it:

> I think it is becoming more and more like that in the last six or seven years. In one sense, it is becoming solely driven by that. How much I talk about it, or if it comes through in the work might vary, and it might reveal itself in different ways, to different people, but that is important. I think there is a beauty in art which offers a different way to reconnect or think about things.

Levene also questions how much the 'artist' should be separated from the 'citizen':

> To practice as an artist is to have privilege, to be free to explore and navigate and present how I see something as a freedom. So, when I began to really engage with the ecological issues that we are all facing it become harder and harder no to use my privilege to share those concerns.

Sometimes the politics are played by the audience. Layla Curtis had an interesting experience when her artwork *United Kingdom* (1999), which was primarily motivated by the opportunity to produce a 'disruptive mapping' piece, was received quite differently by publics in Scotland and England, as she explains:

> That work was first exhibited in Scotland and was seen by a predominantly Scottish audience. At the time the work evoked some strong reactions from those that encountered it – some loved the idea of Scotland as its own separate island, others were offended by the idea of the Scottish landmass being populated by English towns and cities. Later, when the work was included in the exhibition New Contemporaries at the Liverpool Biennial, I sensed the responses the work was evoking were less politically charged. In 1999, Scotland had gone through the process of devolution and the Scottish parliament opened, so Scotland's relationships with its neighbours was on everybody's mind in Scotland. I guess it is not surprising that the work evokes a strong response from those who are directly impacted by the politics it explores.

This political stance of maps relates to the tactics of subversion, as commented on before and was summed up by Curtis:

> You cannot work with maps and cut them up and not involve politics. You are dividing territories and creating new borders, new juxtapositions. [...] The act of cutting a map or somehow re-arranging it feels violent to me and has the potential to create a lot of disruption on the ground.

In the work of Christian Nold the politics of space takes a local tone, as the 'emotional maps' are also highly politicised by some of the observations coming from the public involved, pointing out problems such as inadequate urban design, and unsafe or neglected spaces in the city:

> That is when people start seeing the politics of the place, when they start saying things like 'this traffic here it is really causing a big problem for all of us' or 'it is not just me who feels upset about the fact that I have to wait five minutes to cross the road when I am in a wheelchair and it is very difficult', or 'maybe everybody should do something about this: it is not just my frustration'.

Agency and Transformative Mapping

In an era of extreme visual scrutiny provided by optical remote sensing technologies, from aerial photography to LiDAR (light detection and ranging), and the data attached to them, from where we can extract layers of information, there is still much of the city we do not see. What cannot be seen is not just the cultural layers but also physical layers: what is buried underground, what is behind facades, what is not accessible to most of the public.

As discussed in the previous chapters, counter-mapping initiatives employed strategies to uncover unmapped realities, to lift the veil from hidden territories. As a powerful tool for visualisation, mapping has been used by artists to facilitate the visualisation of information that would be otherwise dismissed or ignored, helping to evidence relationships between physical features and between physical features and social constructs. In her series *Hidden Waters*, for instance, the artist Ruth Levene explored river networks in Britain alongside urban water infrastructures in Yorkshire. The maps were produced while she was in residency at the Faculty of Engineering at the University of Sheffield, and the data was provided by the Yorkshire Water Agency. The two maps of the urbanised areas in Yorkshire, one depicting the water supply pipelines and the other the sewage system, uncover the design of the urban form in a quite unexpected manner, bringing to attention infrastructures that are unseen by citizen dwellers in everyday life, despite their vital, constant use. The map of the river network covering Britain reveals a surprisingly dense pattern, so dense that no other information is added: the rivers themselves draw the recognisable shape of the country. As Levene remarks:

> Mapping is really about revealing something we cannot just see. […] When I first saw the rivers, I was quite physically shocked at how many they were, the landmass that it was. There is nothing there but rivers, but you can recognise the shape of Britain.

The map series enabled the artist to make a political statement about how much of this precious resource had been threatened and how much human life is dependent on it. This revealing aspect, the 'seeing' through the map, is culturally significant in the contemporary world facing the challenges of global warming. The maps are shaped by this historical context, and in turn they acquire relevance as a means to communicate how urban spaces can be thought in relation to the wider environment, bringing aspects of the city that usually are not perceived.

The *Emotion Maps* of Christian Nold take the process of disclosing such unseen realities into the context of urban politics. When mapping collectively, from walking and observing to then sitting down and geo-locating the arousal peaks in a paper map, individuals are brought to see the city from a very different perspective, as they need to not only interact physically with the space but also to talk and articulate ideas about it. People can experience a different way of mapping the city that is not about the demographics, socio-economic clusters or built infrastructure that usually guide urban planning. They are invited to debate the politics of space in practice, as the participatory aspects of the mapping exercises are imbued with a performative aspect. Translating that into a map just enhances the political dimension of the project, pointing to political agency as Nold observes:

> The artistic representation becomes a politics of place [...] you are going the exact opposite away from urban planners, who will start from the Google Earth vision and then drill down and start knocking on peoples' doors. That is why this project has been quite popular, because it triggers people's fantasies for data: planners and politicians do want data, but other people recognise in data human interest and stories. [...] That is why I think it has some political agency in there.

The mapping exercise proposed by Nold is somehow an inverted process of urban planning, 'going the exact opposite way from urban planners', as he affirms. This last observation resonates with the mental mapping of Kevin Lynch, who was also more interested in learning from the public on the streets than from the elaborated planning and design rules decided behind the closed doors of architectural practices and public administrations.

An interesting aspect of Nold's experiments is that the production of a map is not always a necessary outcome, and it is not the focus or main objective. 'The map is a metaphor', as he stated several times during our interview. It does not need to be a mandatory prescription for solving problems: instead, it can be a trope to understand the 'unseen' city, the moods and close perceptions of the urban landscape. In this way participants of the emotion maps are also learning how to see and read the city while walking through it with intention.

Exploring the Urban Pattern

A common characteristic in the production of art maps by contemporary artists is seriality, a strategy that was extensively investigated by conceptual and minimalist artists in the 1960s and 1970s. This is especially true for city maps, where artists tend to produce series based on one specific techniques, such as drawing, GPS tracing, photography, collage and painting, combined with the use of similar formats, defined by size, texture, colours and mode of presentation. The recognisable patterns that artists aim to imprint in their works relate to investigations of urban phenomena, from analysing their physical appearance and intrinsic design to perceptions acquired by means of personal experimentation. Even if these maps suggest regularity and order at first glance, they should not be mistaken as mere exercises of representational techniques. In this section I present and comment upon the serialised artworks of five artists.

While unique and individual pieces, maps composed within series disguise a similarity that at first sight appears uniform but in fact gives an entirely different meaning to the artwork regarding the unequivocal repetition of patterns. They are detectable from the design created by urbanisation viewed from above, grid-like shapes intersected by major lines and polygons of open spaces. Urban art map series denote a play with the figurative aspect of the urban landscape that often is intensively re-worked by the artists, overlapped, redrawn, embellished, dismantled, decomposed and rearranged. This 'over-mapping', reinforced by the serial presentation, demands from the observer an investigative eye in order to find rhythms and their breaks, similarities and discrepancies. In common these artworks are cleared of any cartographic codes and conventions: none of them present a legend or key, or indications of any nature about scale, projection, orientation, place names or topographical features. The city names on the titles are the only clue to their location on the globe. In doing so the artist can engage with the primordial aspects of the urban form, stripped of categories, adjectives, fantasy or politics: they present 'raw' portraits of cities; yet they are manifestations of very personal interpretations of urban life and landscape.

The first artwork to be commented is the *City Drawings*, which artist Kathy Prendergast started as a series in 1992 attempting to produce maps of all the world's capital cities.[12] The maps are hand drawn in pencil on A4 sheets, composed by the fine tracing of lines on paper, rendered as a delicate lace work with intricate texture. In an interview with Coline Milliard in 2009, Prendergast talks of her work with maps as a quest to explore geographies and personal identities. In her *City Drawings* this relationship is posed by a metaphor that connects city shapes to human features:

[12] Between December 1999 and March 2000, the Irish Museum of Modern Art exhibited 113 completed drawings. McKee, F. and Irish Museum of Modern Art (1999). *Kathy Prendergast: the end and the beginning*. Exhibition catalogue. London: Merrell.

> I started to look at maps of cities to see if you could guess how they evolved just by looking at them, a bit like following people's footprints [...] I realized that all cities do grow from the inside out [...] as if a collective consciousness was forming the city [...] I drew my maps the size of human faces to comply with that idea of a portrait, so we could see them as a reflection of ourselves [...] I would try and look at the map as if the roads had just grown out of people walking.[13]

Prendergast states that her work is not documental and does not involve historical research; the map is the sole source of information. Questioned on how she places her maps in relation to 'real maps', she replied that:

> I do not see them as maps, I see them as drawings of cities.[14]

One could argue that, inversely, 'real maps' can be seen as drawings that 'real' urban cartography appeals to the viewer's eye as an aesthetic artefact. In producing a massive volume of the drawings and exhibiting them in their 'naked' form, with no frames, just placed side by side,[15] Prendergast plays with the artificial nature of cartography, through investigating what shapes the urban pattern.

The second artwork is the series *My Ghost*, where the artist Jeremy Wood registered his daily movements in London, where he lives and works, during a period of 16 years, between 2000 and 2016. Wood initiated his GPS drawings in 2000 to 'explore the expressive qualities of using his body as a geodesic pencil', as stated in the artist's website.[16] Wood creates his maps carrying a GPS in his journeys, at random or intentionally, everywhere he goes. In an interview given to Tracey Lauriault in 2009, the artist talks about his impulse to map his own movements:

> What strikes me about seeing one's own tracks is in the triggering of memories, which is why I refer to them as cartographic journals. They are a record of where I have been and a daunting reminder of where I have yet to go. Among the intricacies in the line qualities, I can read my routes and ruts,

[13] Millard, Coline (2009). Kathy Prendergast in interview. *Circa*, n. 128, pp. 42–51.

[14] Ibid., p. 44.

[15] As it was exhibited in the Venice Biennale in 1995, when she won the prize for best young artist.

[16] See https://www.jeremywood.net/about.html.

> and recall my dithering and my adventures. They are of my ghost, captured in places of a different time.[17]

For the *My Ghost* series the artist, as composing a visual diary, registered in 'notebooks', in this case, printing maps, his 'past' self through annotation of the lines resulted by his dislocations. The ones produced in the years of 2009 and 2015 are particularly suggestive: the prints of striking white lines against a dark background recall 'apparitions' and express movement as thick, more defined lines contrast with the fine and fainted ones, like vapour trails left by the passage of aircraft in the sky.

The third artwork is the *Diorama Maps*, produced by Japanese artist Sohei Nishino, who started the series in 2003 photographing Osaka. His method has been the same since then: camera in hand Nishino goes walking around the city capturing panoramic views and fragments of the landscape, whether entire buildings, blank skies, water surfaces or stains in the pavements. Back in his studio he develops and prints thousands of photographs (he works with 35 mm film), arranging them manually over a canvas to compose a new picture, a city map of his own, which is then printed and framed. The series so far includes 20 dioramas of 19 cities (he mapped Tokyo twice, in 2004 and 2014) around the world[18]; the final prints measure between 1.3 × 1.2 m (Osaka) and 2.6 × 1.8 m (Johannesburg) and have been exhibited widely in Japan, the USA and Europe.

Nishino goes on his wanders with a city guide, but what results from this long and laborious process is an entirely different product that can be more related to a personal insight, a portrait made of infinite small details of a place. As he states:

> [...] the finished work is anything but an accurate map, it is simply the town as seen through the eyes of a single individual, an embodiment of my awareness, a microcosm of the life and energy that comprises the city.[19]

[17] Lauriault, Tracey and Wood, Jeremy (2009). GPS tracings—personal cartographies. *The Cartographic Journal*, 46, 4, pp. 360–365.

[18] The cities are, in chronological order of the creation of the diorama, Osaka (2003), Hiroshima (2003), Kyoto (2003), Tokyo (2004), Shanghai (2004), New York (2006), Paris (2008), Hong Kong (2010), London (2010), Istanbul (2011), Rio de Janeiro (2011), Berlin (2011), Bern (2012), Jerusalem (2013), New Delhi (2013), Tokyo (2014), Amsterdam (2014), Johannesburg (2015), Havana (2015) and San Francisco (2016).

[19] Artist's statement. Available at http://soheinishino.net/dioramamap.

What makes his work singular is a clever combination of the view 'from above' with the view 'from the ground'. Nishino re-enacts the 'God' map maker, surveying everything and everybody with the detached eye of a superior entity, while at the same time he plays the pedestrian role, mingling among the crowds, interfering with their movement, taking part in the big picture.

Next is the artwork of French artist Armelle Caron, who produced her first *ville rangée*[20] in 2005, while living in Berlin. This initial experiment developed into the series *Les Villes Rangées*, which explores the figure/ground typology of ordinary maps, playing with solid shapes and re-arranging them in order to create a puzzle—which is then de-puzzled. What does this new typology represent—is it still a city? This game of manipulation and displacement, as in the situationist map, compels the reader or observer to reflect on traditional concepts of urban hierarchies and structures.

Regrouping the shapes of built blocks and public open spaces, Caron's maps work as an anagram game that is played with a visual alphabet, each time resulting in a new script. The linear composition of forms suggests the lines of a text that can be 'read' as a musical score with a rhythmic scale, each block representing a single note. The artist considers her work as a 'personal archaeology', which she explores in her own way:

> My personal story as a traveller binds me to the maps; I sometimes look at them as the image of a territory and sometimes as a graphic image empty of any representation. That is why in this work I removed the captions, the colours, the textual elements. [...] I seek a visual order that pleases me, that reminds me of the written word, intending that this image tells me something else than the city itself.[21]

The last example is Damien Hirst's series of 17 paintings entitled *Black Scalpel Cityscapes.* Produced in 2014 these are bird's-eye views of cities made up of 'scalpel blades, skin graft blades, zips, stitching needles, aluminium filings, pins, stainless steel studs, fishhooks, steel wire cutting spool and gloss paint on canvas, against a dark background'.[22] The listing

[20] Translated as 'city ranked' or aligned, in allusion to the linear form of the shapes.

[21] Interview for the magazine ENTRE n. 10, mai 2014. Available at http://www.armellecaron.fr/textes/.

[22] According to the exhibition catalogue's description. Hirst, Damien; Bracewell, Michael; Brotton, Jerry and White Cube Gallery (2014). *Black scalpel cityscapes—Damien Hirst.* London: Other Criteria.

of the materials that accompanies the works gives a quite harsh first impression; we are immediately warned that these are not conventional pictures.

The artist described them as 'portraits of living cities', alluding to contemporary urban themes of physical transformation, surveillance, globalisation and conflict (most of the cities selected are either major conurbations or sites of religious and political dispute). The use of surgical instruments suggests ambiguous meanings, evoking the violence experienced in large cities at the same time as the brightness and light, reflected by the precision crafted metal, that attract people to the same places.

As a cityscape viewed from above, the maps' subjects are distinguished by the intricacy of their thoroughfares, and part of the playfulness in these art maps series rests on reducing the urban network to its bare elements, such as the tracing of lines in Prendergast's and Wood's works, or completely ignoring it, as in Caron's lined shapes and Nishino's collaged snapshots. On the other hand, the profusion of imagery in Nishino's dioramas or the over-elaboration of the 'sculpted' paintings of Hirst's pieces takes the observer into a quasi-optical illusion of a three-dimensional space. These five series of city maps encourage an interrogation about those visual relationships that underpin geographical knowledge; they might look like maps but certainly they do not function as guides for navigation or place finding. Instead, the knowledge they provide is of that triggered by imagination, of the artist and of the viewer as well.

In the light of what the interviewed artists reveal through their statements about the process and intentions behind each work, it becomes clearer that their mapping exercises demand a more personal connection with spaces, and much of their craft involves 'transferring' these connections, emotions and reasonings into the artwork. They are also preoccupied in finding means to communicate their ideas, or to send a message, to the viewer; this is done by breaking cartographic conventions, demystifying 'accuracy' and 'precision', questioning the use and purpose of maps, revealing other aesthetic values in visual representations, bringing the 'lived space' to the foreground as well as enhancing the importance of individual experiences. The next chapter presents the case study of Liverpool, with an introduction to its historical urban development and its corresponding cartography, and to then examine three contemporary art maps through interviews with their authors.

CHAPTER 6

Liverpool: A City in Maps

Liverpool is probably most famous for being the birthplace of the Beatles, the rock band that became a global icon of popular culture in the 1960s and made the city an international tourist destination for music fans. But the streets and buildings depicted in souvenirs and forming the background to tourists' selfies carry many other stories, of a place of power and wealth, deprivation and struggle, pride and hope. The following section presents a brief history of Liverpool's development as a city along with its cartographic representations, explaining the research context that informed the artists who created the three contemporary art maps of Liverpool between 2008 and 2011. These are analysed in the second part of this chapter, together with the main topics discussed during the interviews. The format of these interviews is similar to the ones presented in the previous chapter: informal, long talks in the artists' studios (in London, with Stephen Walter and Ben Johnson, and in the National Glass Centre, Sunderland, with Inge Panneels and Jeffrey Sarmiento), and the same questions relating to their practices and methods were asked, with a particular focus on the Liverpool maps.

The history of Liverpool's cartography, as of any other city, reflects the importance of registering a place to reveal its past layers of social and economic history. Whenever maps are produced, they leave a trace, impregnating the collective memories of city dwellers with strong images that persist to present times, and the feelings and emotions about 'the past' that are perpetuated in such documents certainly influence perceptions

G. Lanci, *Art Maps and Cities*,
https://doi.org/10.1007/978-3-031-13306-0_6

about a city. Liverpool today is one of the major conurbations in the northwest region of England,[1] but from its foundation as a borough in 1207 to the mid-1600s, when the first harbour was built (1635), Liverpool remained a rural fishing village with insignificant population growth. It was only in the early eighteenth century, with the expansion of Britain's lucrative trans-Atlantic trade, in association with the industrial revolution that dramatically changed the country's landscape, that Liverpool was catapulted into exponential urbanisation and its territory was extensively mapped and documented.

The borough of Liverpool was officially designated in a Royal Charter of August 1207, signed by King John. There are few records of Liverpool's existence before that: there are no remains of a Roman settlement, and the city is not mentioned in the Domesday survey of 1086. However, there is no indication that after 1207 much happened, and the oldest known map in existence, dated 1572, depicts the same streets that originated the settlement, in an 'H' shape, and the Castle, the only significant building, erected in the 1230s and demolished in 1715. A painting by an unknown artist, belonging to the Merseyside Maritime Museum collection and commonly referred to as 'Peters painting', shows the Liverpool waterfront in 1680.[2]

The painting takes a viewpoint, looking at the city from the River Mersey, showing no more than a few prominent buildings (the old town hall, St. Nicholas Chapel, the Custom's House and the Castle), one or two warehouses and minor dwellings. But in the foreground, the Mersey is packed with a great variety of vessels, a sign of the prominence that shipping and maritime trade will have in the centuries to come in Liverpool's history.

The 'Peters painting' also served as the base for what might be the first bird's-eye view of the city, the *West prospect of the town of Liverpool as it appeared about the year 1680*, as drawn by John Eyes and engraved by

[1] The city's last estimated population by mid-2020 is 500,500 in the metropolitan district (local authority) and 1,434,300 for the Merseyside County region, being considered the fifth-largest conurbation in the UK. Source: Office of National Statistics, Population estimates in the UK, 2020. Available at https://www.ons.gov.uk/peoplepopulationandcommunity/populationandmigration/populationestimates#datasets.

[2] The painting was presented to the Liverpool town council in 1818 by Ralph Peters III, but it was probably in the family's possession since the 1700s. For Tibbles, the Peters painting is a 'near contemporary depiction of Liverpool about 1680' with accurate topography and shipping. Tibbles, Anthony (1996). Two paintings of Liverpool in 1680: a reassessment. *Transactions.* The Historic Society of Lancashire and Cheshire. Vol 146, pp. 189–195.

B. Howlett. John Eyes was a land surveyor and produced various charts of the English coast and maps of Liverpool between 1738 and 1772 (Fig. 6.1).[3]

After his interpretation of the 1680s' panorama of Liverpool, John Eyes produced surveyed maps that documented the transformations in Liverpool during the eighteenth century, as in his *A Plan of Liverpool as it was in the Year 1725*. He was not alone, and other surveyors joined him in the task of mapping the rapid changes in Liverpool. In 1725 James Chadwick produced the first official street map, *The Mapp of all the Streets, Lanes & Alleys within the Town of Liverpool with the side of the River Mersey layd down by a scale of 80 y[ar]ds in an inch* (Fig. 6.2). The map is composed using rich imagery and textual information, with particular attention given to the detailed list of the names of 60 streets, laid out in two columns siding the central map. The long title reinforces the scientific accuracy of the work, indicating that this is a city of importance.

In both maps there is a significant feature: 'The Dock', as labelled in Eyes' map. It was inaugurated in 1715, being the first commercial wet dock in the world. The maps also show the 'pool' that has given the name to the city filled in, streets laid out and land ready for urban expansion. These maps are a testimony of a crucial time in the history of Liverpool, when its whole landscape was being radically transformed, putting the city on the path to become one of the wealthiest places in Europe, through its involvement with the Atlantic slave trade and its role supporting the cotton spinning industry of northern England.

Britain began trading with Africa in the mid-sixteenth century—the first recorded ship set sail to the Guinea and Benin in 1553—and for a long period, until 1698, London enjoyed an exclusive monopoly on the African slave trade. With the opening of the trade to private merchants, the city of Bristol soon took the lead in the business, while Liverpool started slowly, dispatching only two slaving vessels in the first decade of the eighteenth century.[4] However, by the 1740s, Liverpool had caught up with London and Bristol, and by the 1750s more than a half of all vessels

[3] According to existent registers at Liverpool Record Office archives. Available at http://archive.liverpool.gov.uk (see results consulted under the tag 'John Eyes').

[4] The London-based Royal African Company's monopoly in the slave trade officially ended in 1698 but much before that Bristol was sharing the market with London and by the 1720s became one of the wealthiest cities in England. Morgan, Kenneth. Liverpool's dominance in the British Slave Trade, 1740–1807.Richardson, D., Schwarz, S. and Tibbles, A. (2007). *Liverpool and the Transatlantic Slavery*. Liverpool: Liverpool University Press. pp. 14–42.

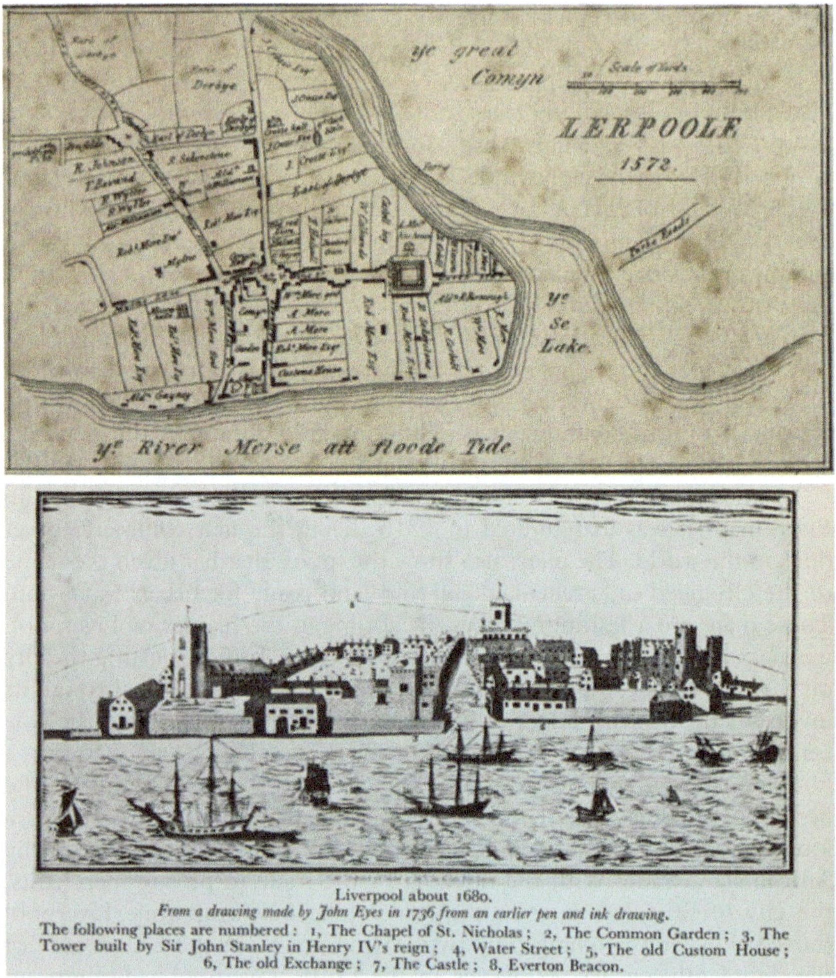

Fig. 6.1 Liverpool as would be in 1572 (above), from an unknown author, and in 1680 as depicted in John Eyes' engraving from 1736, based on the 'Peters painting' (below). Source: Liverpool Record Office

departing from Britain to Africa were sailing from its shores. From then on, the merchants of Liverpool dominated the slave trade until it was ended in Britain in 1807.[5]

[5] Morgan, Kenneth (2007). Op. cit., p. 15.

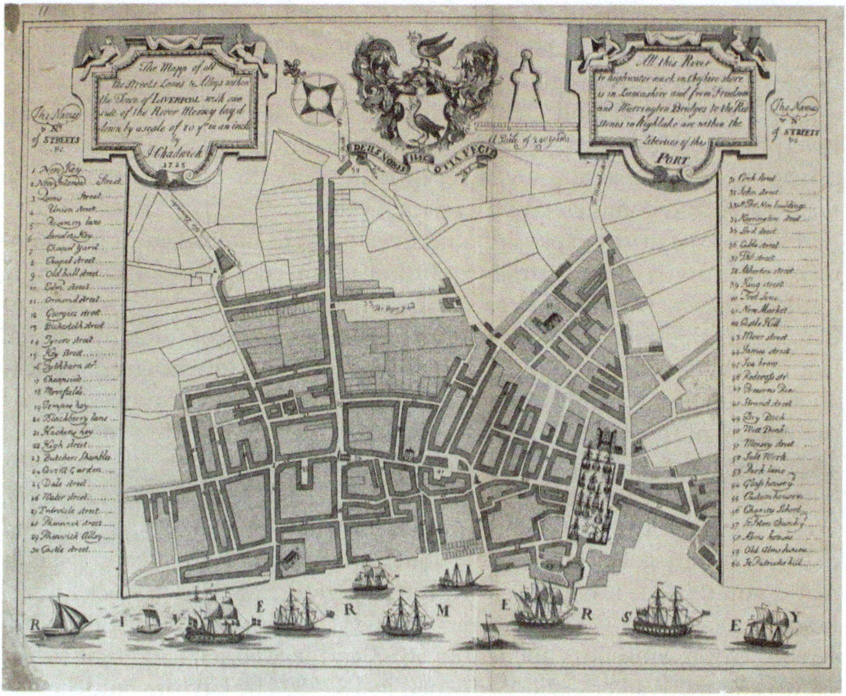

Fig. 6.2 James Chadwick's Street Plan of 1725. Source: Liverpool Record Office

The surveyor John Eyes went on to produce another four town plans, in 1753, 1765, 1768 and 1772. Following his uncle's steps Charles Eyes took over the surveying office and authored the plans of 1785 and 1796.[6] The cartographic record of Liverpool in the eighteenth century points to an increased importance of the city in the national and international economic scene, documenting its rapid growth in wealth and prosperity. Looking at John Eyes' plans from 1725 to 1796, one can clearly observe the transformation of an ever-growing urban area and in particular the improvements in the docklands. The former monumentality of the Old Dock was very soon supplanted by even greater system of dry and wet docks along the Mersey to accommodate the rising volume of the port's shipments.

[6] Op. cit. Liverpool Record Office archives.

Liverpool's population rose from about 7000 inhabitants in 1708 to 34,000 in 1773 and reached to over 77,000 by 1801.[7] The industrialisation of the northwest region, mostly related to textiles, expanded greatly during that period and benefited from the demographic growth of Liverpool and the development of its port infrastructure. Liverpool's proximity to manufacturing centres in Lancashire turned the city into the main social and commerce hub outside London.

The pace of urban expansion and development intensified in the nineteenth century as the industrial revolution advanced. Liverpool had already in place the financial resources and qualified labour force to continue to develop its shipping business, and increasing numbers of migrants, casual labourers and seafarers populated the waterfront, making the city a sought-after destination to live and work. The city enjoyed its economic high by the middle of the nineteenth century and established itself as a major port attracting thousands of immigrants: indeed, by the 1880s, the city was known as 'the New York of Europe' and 'Second City of the Empire'.[8] Wealthy merchants invested heavily in infrastructure in the docks area, and many of Liverpool's impressive public buildings and architectural landmarks still present today come from this period: the Albert Docks warehouses (1846), the Central Library (1852), St George's Hall (opened in 1854), World Museum (opened in 1853 and moved to its present site in 1860) and the Walker Gallery (opened in 1877). The city was embellished with designed botanic gardens and parks, such as the Stanley Park, opened in 1870, and Sefton Park, which followed in 1872.

Many of the greatest engineering achievements of the Victorian period were introduced in Liverpool: the first air conditioning system in a public building in the UK, installed in the St George's Hall by the time of its construction, the first intercity railway connection in the world (Manchester-Liverpool, 1830), the world's oldest underground railway outside London (Mersey Tunnel, 1886) and the world's first elevated electric railway (Liverpool Overhead Railway, 1893). Urban growth was gaining prominence, and Liverpool started to absorb surrounding villages and districts, slowly taking the shape of an urban conurbation.

[7] Morgan, Kenneth (2007). Op. cit., p. 17.

[8] Belchem, John (2006). *Liverpool 800: culture, character & history*. Liverpool: Liverpool University Press.

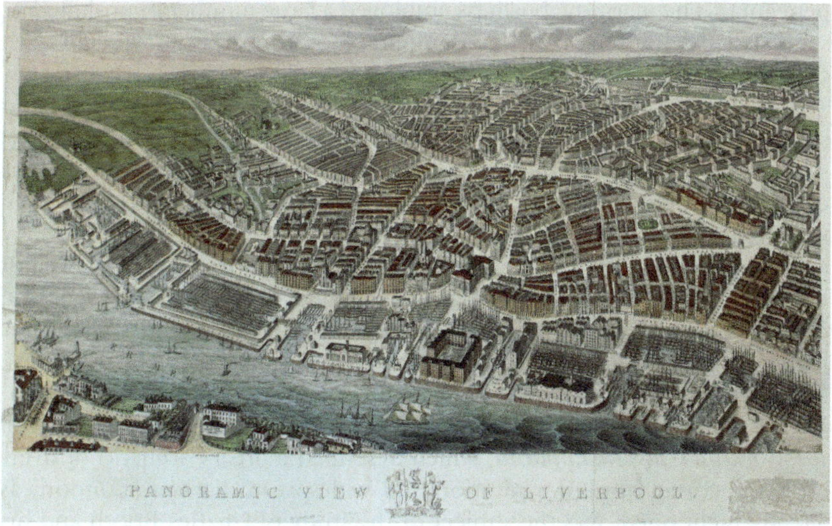

Fig. 6.3 View of Liverpool from the Mersey. Rudolph Ackermann, 1847. Source Liverpool Record Office

The waterfront was constantly changing, the docks becoming ever bigger and dominant in the landscape. Prince's Dock opened in 1821; the Old Dock was filled in 1826 and a new Custom House was built on the site in 1839; and Albert Dock was inaugurated in 1846. In Rudolph Ackerman's magnificent birds'-eye view of the city of 1847, the complex of harbourage and shipping yards is enhanced with hundreds of sailing masts, covering every single slot at the docks. The map is a statement of the grandeur of the emerging metropolis, probably one of the last of its kind before topographical maps derived from accurate land surveys took over (Fig. 6.3).

The city's population grew exponentially, in tandem with the industrial revolution and commercial expansion. Liverpool was acquiring a status of regional capital, and as a city of business opportunities and employment, it attracted migrants and casual labourers from all parts of the UK and, mostly, Ireland. In 1851 almost half of Liverpool's population was born outside Lancashire, the Irish-born being the most significant group (22.3

per cent).[9] The maritime business employed a large number of unskilled or semi-skilled labourers, and this demographic group was also linked with chronic social problems, such as disease and violence.

Liverpool in the Victorian period turned into a city of splendour and squalor. Poverty and class conflict, with acute social and economic divisions, were reflected in the urban landscape of segregated housing, concentrating working-class areas in slums housing close to the docks.[10] Crime was rife in those areas and beyond; some criminals were driven by poverty and need but others by the many opportunities that economic wealth brought for gambling scams, racketeering and other types of organised crime.[11] Urban poverty and social struggle were characteristics not peculiar only to Liverpool but could be observed in other major British cities. Manchester and the northern industrial towns, for instance, were the subject of the classic study conducted by Friedrich Engels, *The Condition of the Working Class in England*, which described a dreary scenario of overcrowded cottages, poor sanitation and widespread diseases. London, as already mentioned, was the subject of the study of Charles Booth and his famous map.[12]

Maps of Liverpool from the second half of the nineteenth century were more committed to the accuracy of land surveying; the pictorial elements—the facades of important buildings and the fleet of ships crossing the Mersey—were all but ditched in favour of information: lists of street names, names of the main public buildings and parks, census population, travel distances and scale (Fig. 6.4). They also gradually abandoned the North-South horizontal layout, which enhanced the waterfront landscape and the city centre, to adopt the vertical, North up orientation. At the

[9] It must be noticed that there was a major emigration of Irish worldwide between 1845 and 1849, during the Great Famine. Lawton, Richard. The components of demographic change in a rapidly growing port city: the case of Liverpool in the nineteenth century. In: Lawton, Richard; Lee, Robert (eds.) (2001). *Population and Society in Western European Port-Cities, c. 1650–1939*. Liverpool: Liverpool University Press. (pp. 91–123; p. 115).

[10] '[…] public health officials and social investigators had no hesitation in applying an ethnic label to the worst areas of the city, the 'squalid Liverpool', running along the low-lying westernmost strip from the Dingle to Sandhills'. Belchem, John (2007). *Irish, Catholic and Scouse: the history of the Liverpool-Irish, 1800–1939*, Liverpool: Liverpool University Press. (p. 57).

[11] Macilwee, Michael. (2011). *The Liverpool Underworld: Crime in the City, 1750–1900*. Liverpool: Liverpool University Press.

[12] Humpherys, Anne. (2002). Knowing the Victorian City: Writing and Representation. *Victorian Literature and Culture, 30*(2), 601–612.

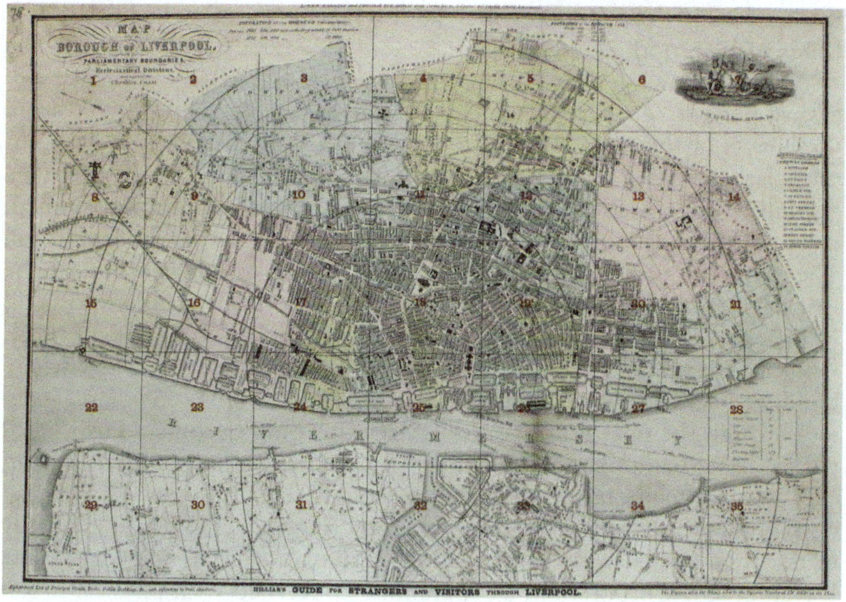

Fig. 6.4 Map of the borough of Liverpool with the parliamentary boundaries, ecclesiastical divisions and a part of the Cheshire Coast: Hilliar's Guide for strangers and visitors through Liverpool, 1858. Source: Liverpool Record Office

dawn of the twentieth century Liverpool maps are far more numerous and sophisticated; the cartographic documentation kept in pace with other great British and European cities at the time. As the city grows, maps from the 1900s onwards show a Merseyside region centred on Liverpool, connected to Birkenhead by steam ferry and underwater tunnel, and to the hinterland by a railway network.

At the turn of the twentieth century a set of three new buildings would imprint in the city its most remarkable skyline: the Port of Liverpool Building (1907), the Royal Liver Building (1911) and the Cunard Building (1917). Known as the 'Three Graces', these buildings remain to the present the landscape signature of Liverpool, perpetually illustrating tourist brochures and posters. The city's population grew exponentially with almost no interruption between 1801 (78,000) and the mid-1930s (870,000), when it reached its peak.[13] Then, the fortunes of Liverpool changed again.

As the twentieth century unfolded, Liverpool started to experience successive periods of social struggle and deprivation. The city suffered from the effects of the Great Depression of the 1930s that damaged its economy, largely based on the port area and related activities; during the Second World War, the central and port areas were heavily bombed, demanding a great effort of social, physical and economic reconstruction. The crisis intensified from the 1960s onwards with the decline of the British manufacturing industry and competition with other ports in Europe, resulting in economic depression that inflicted the loss of thousands of jobs and a sharp decrease in population of the inner city, where buildings were left to decay.

Through the next four decades Liverpool would become a busy 'laboratory' for urban policy and planning in the UK, hosting numerous projects on social housing, education, commerce, business and urban development, in order to recover the city.[14] Regeneration began gradually in the mid-1980s, with governmental initiatives such as the Merseyside Development Corporation, a programme concerned with physical urban development that allowed the recovery and renovation of the Albert Dock between 1986 and 1988, among other works. In 1994 the city council gained substantial financial aid from the European Union funding through Objective 1, a programme designed to help impoverished urban areas,[15] and when the New Labour government launched its Urban Task Force in 1998, a programme focused on tackling urban decline through urban design initiatives, Liverpool was the first city in the UK to establish its own Urban Regeneration Company, Liverpool Vision, encouraging developments in and around the city centre.[16]

Since the 1980s regeneration projects in Liverpool gave a considerable boost to the cultural sector, which was viewed as a strategy to stimulate

[13] Sykes, Oliver et al. (2013). A city profile of Liverpool. *Cities*, 35, pp. 299–318 (see Graph 1, p. 300).

[14] Couch, Chris (2003). City of change and challenge: urban planning and regeneration in Liverpool. Aldershot: Ashgate.

[15] The funding was secured for 12 years, between 1994 and 2006. Sykes, Oliver et al. (2013). Op. cit., pp. 310–311.

[16] The company has now been closed since November 2018. Liverpool City Council announces the closure of Liverpool Vision. Liverpool Business News, November 22. https://lbndaily.co.uk/liverpool-city-council-announces-closure-liverpool-vision/.

tourism in the city and attract new private investors. From that time two events that had a great impact in the city's cultural life were the International Garden Festival (1984), which attracted more than three million visitors, and the inauguration of Tate Liverpool (1988) at the Albert Dock, the first Tate gallery opened outside London. Afterwards a succession of initiatives continued to target cultural venues around the city centre, such as the refurbishment of the Royal Philharmonic Hall, home of the oldest professional symphony orchestra in the UK, in 1995. When Liverpool Vision came into effect, areas such as Ropewalks, a neglected working-class neighbourhood, were targeted for redevelopment of public spaces and redirected to be transformed into a cultural and leisure quarter.[17]

At the dawn of the twenty-first century, the negative image of Liverpool as a city of economic and social decline was about to change. In June 2003 Liverpool was announced as the winner of the bid to host the European Capital of Culture (ECC) for 2008; in 2004 UNESCO designated six areas in the historic centre and the docklands of the maritime mercantile City of Liverpool as a World Heritage site, and in 2007 the city celebrated its 800th anniversary with events spread citywide, in anticipation of the Capital of Culture year. The city's 'renaissance' was enhanced as well by the Liverpool F.C. team in the international and national scenes: they were European champions in 2005 (over 20 years since the last title) and winners of the FA Cup in 2006.

The ECC designation prompted a wave of urban developments in the cultural and business sectors alike, supported by Liverpool Vision. In 2003 the city received a new cultural venue, FACT (Foundation for Art and Creative Technology), established right in the middle of Ropewalks and the Bluecoat Arts Centre, which was reopened in 2008 after three years of renovation works and the construction of a new extension. In addition, in 2008 two major developments were inaugurated in the city centre alongside the waterfront: Liverpool One, a retail development that combines shops, residential units and leisure facilities, and the Liverpool Arena, an events and conference venue. Other redevelopments next to the city centre include the renovation of the Baltic Triangle area, which started in 2010 and is ongoing, and the revamp of the waterfront with a mix of new buildings, including the Liverpool Cruise Terminal, opened in 2007, and the Museum of Liverpool and adjacent complex of the Mann Island

[17] Bayley, Stephen and Liverpool City Council (2010). *Liverpool: shaping the city.* London: RIBA.

buildings, all inaugurated in 2011. The revitalisation of the city centre brought the population back to the area, which quadrupled in two decades, from the few thousands in the early 1990s to 36,000 in 2012.[18] More recently new residential developments have appeared in the central area, especially those built for student accommodation, although not without criticism.[19]

The legacy of the urban regeneration projects and the ECC remains present in the city, but the benefits are not evenly spread around the whole of the urbanised area. Deprivation persists across Liverpool, and poor enclaves, with derelict housing estates, can still be found even in the city's inner core.[20] The politics of austerity implemented in 2010 certainly did not help to keep the momentum of the 2000s renaissance, although the positive image built since then contributed to create a more resilient atmosphere in the city (Fig. 6.5).

Contemporary Art Maps of Liverpool

In 2008 the city of Liverpool was about to stage the European Capital of Culture year, and a map commissioned in 2005 was ready to be exhibited to the public. Its creator, Ben Johnson, was the artist invited for the task, being known for large-scale paintings of architectural spaces, depicted in great detail. His website provides videos from which one can gain a good insight into his meticulous creative process. Johnson describes himself as a 'perfectionist' with an 'obsessive' working practice; usually his paintings take years to be completed (Fig. 6.6).

Ben Johnson was born in Llandudno, Wales, in 1946, and studied at the Royal College of Art in London, to where he moved in 1965 and

[18] Sykes, Oliver et al. (2013). Op. cit., p. 312. According to Liverpool City Council key statistics, the population in 2017 of the enlarged central area, comprising the Central, Riverside and Prince's Park wards, was 72,200. Available at https://liverpool.gov.uk/council/key-statistics-and-data/data/population/ (Table p4: 2017 Ward Population).

[19] Mulhearn, Chris and Franco, Michael (2018). If you build it, will they come? The boom in purpose-built student accommodation in central Liverpool: destudentification, studentification and the future of the city. *Local Economy: The Journal of the Local Economy Policy Unit*, vol. 33, no. 5, pp. 477–495.

[20] In 2015 Liverpool ranked the fourth most deprived local authority area in England; this can be considered an improvement as the city ranked first in the Index of Multiple Deprivation in 2004, 2007 and 2010. Liverpool City Council (2015). *The Index of Multiple Deprivation 2015: a Liverpool analysis*. LCC, 2015. Available at https://liverpool.gov.uk/council/key-statistics-and-data/indices-of-deprivation/.

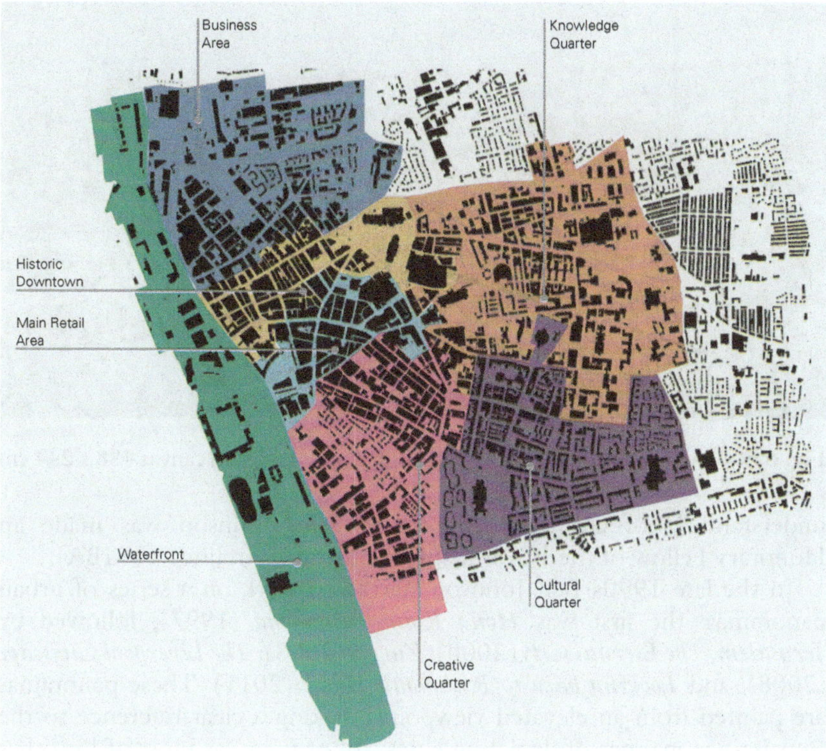

Fig. 6.5 The urban mosaic: the map shows Liverpool city centre dived into areas that are framed around major urban regeneration projects and key buildings and institutions. The 'Main Retail Area' corresponds to Liverpool One; the 'Creative Quarter comprises Ropewalks and Baltic Triangle; the 'Cultural Quarter' contains the two cathedrals and a series of cultural venues, such as the Liverpool Institute for Performing Arts, the Royal Philharmonic and the Everyman Theatre; and the 'Knowledge Quarter' includes University of Liverpool and Liverpool John Moores University campuses. Source: *Liverpool City Centre Strategic Investment Framework*, 2012. (Liverpool Record Office)

established his studio. His first solo exhibition, with works inspired by German expressionism, was held in New York in 1969, and it was a commercial success, allowing him to set up an art studio in London. Johnson soon abandoned his expressionist paintings and made architecture his central subject; in 1995, for his contribution to the promotion and

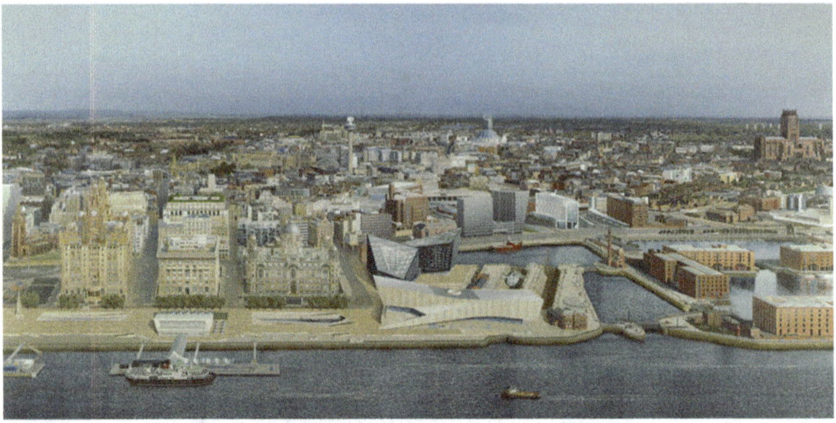

Fig. 6.6 The *Liverpool Cityscape* by Ben Johnson. Acrylic on canvas 488 x 244 cm

understanding of contemporary architecture, Johnson was made an Honorary Fellow of the Royal Institute of British Architects (RIBA).

In the late 1990s Ben Johnson started to work on a series of urban panoramas: the first was *Hong Kong Panorama* (1997); followed by *Jerusalem, The Eternal City* (2000); *Zurich* (2003); *The Liverpool Cityscape* (2008); and *Looking back to Richmond House* (2011). These panoramas are painted from an elevated viewpoint, making a clear reference to the Renaissance manner of depicting cities. All of them, as most of his other painting works, have been commissioned and remain in private collections, except for the Liverpool Cityscape, which is in permanent exhibition at the Museum of Liverpool.

Ben Johnson's *Liverpool Cityscape* is a large painting, almost two and a half meters tall and five meters long, that depicts a classical bird's-eye view of Liverpool city centre, its most iconic buildings and its surroundings. For three years, the artist and five assistants worked full time in the studio, along with 43 people who collaborated intermittently; more than 3000 photographs were taken, almost 23 stencils were used and 700 individual colours were mixed.[21] The production process is mostly based on photographic documentation taken by the artist in the city, which was then used

[21] Bukantas, Ann (ed) (2008). *Cityscape: Ben Johnson's Liverpool*. Liverpool: Liverpool University Press.

for creating computer-generated drawings from stencils that were produced to apply the painting on the canvas.

This intriguing work seems to have two distinctive natures: a very formal one, where the production process is committed to detail and perfection, and a fantasy-like one. In the related video posted on his website, Johnson comments: 'What I'm painting is perhaps a fantasy of the city. It is a city seen at a great distance. [...] There are no people, there are no cars. There is an invitation for somebody to enter that dreamlike world and inhabit it'.[22] After three years of intense work, in 2008, the painting was moved from the artist's studio in London to be finished at the Walker Art Gallery in Liverpool, during a six-week residency, under public view. According to the artist, 'it was always my intention to finish it in public so that people could see that the work didn't come out of an ivory tower; it's actually made by very ordinary people'.[23] The painting remained in the Walker Gallery until 2011, when it was moved to the newly inaugurated Museum of Liverpool.

While Ben Johnson's map was being inaugurated, another artist was wandering the streets of the city registering his experience on a map. Stephen Walter's *Liverpool 2008–09* is a hand-drawn map made within a period of six months. From his own account during his initial visit to the city, he was sent on a wild goose chase by The Guardian newspaper to make a map of the city, but this project fell through. It was on this visit that he met and fell in love with a woman from Liverpool, and this chance meeting led him to further explore the local history and people, resulting in the making of a map 'far larger and more detailed than was initially planned'.[24]

Steven Walter grew up in Barnet, on the outskirts of London, being born there in 1975. Walter received his degree in Fine Arts from Manchester Metropolitan University in 1998 and went to complete a MA Fine Art Printmaking in 2001 in the Royal College of Art in London, where he still lives and works. Walter's main body of work consists of intricate hand drawings that are inspired by observation and experience of landscapes. He had started drawing maps in 2002, crowding the surface with symbols and written text that combine folklore, historical facts and his own personal views. Stephen Walter's map of London entitled *The*

[22] See www.benjohnsonartist.com/cityscapes-2/2015/8/28/the-liverpool-cityscape.

[23] Ibid.

[24] See www.stephenwalter.co.uk/the-maps.

Island (2008) figured in the major exhibition held at the British Library in 2010 'Magnificent Maps', and in 2012, he was commissioned by the London Transport Museum to create *London Subterranea*, a map of the city's underground landscape, to take part in the museum's group exhibition 'Mind the Map'. Walter's work can be also found in permanent collections of the British Museum, the Victoria and Albert Museum and the Courtauld Institute of Art in London.

The map of Liverpool is a product of the first-hand experience of the artist in the city, web-based research and historical maps and books. Public figures, historical events, folklore, built infrastructure, pubs, football stadia and green spaces were placed accurately, using an up-to-date Ordnance Survey map as cartographic base. 'I began to translate Liverpool's character on to paper, its quirks, idiosyncrasies and its stereotypes into a celebration of the masonry and hearts of Liverpool'.[25] The map is drawn in scale, as a topographical map, and it covers the whole extension of the urbanised area. On the top right corner there is an extensive legend ('key' on the map) and at the bottom left an inset displaying the city centre in detail. The array of information is vast and showed in what may appear at first sight to be a cacophony of notations, but looking closer the map reveals more of its structure. It depicts a preference for the picturesque and curious, political commentary and humour. The legend is suggestive of the mix of historical and geographical survey, anecdotal facts and personal observation of the urban landscape. The map is in permanent exhibition at the Bluecoat Art Centre in Liverpool (Fig. 6.7).

At the same time that Johnson's painting was completed, and Walter started his hand-drawn work, a national competition was launched in June 2008, asking for a cultural map of Liverpool to celebrate its European Capital of Culture year. The map would be placed in the new Museum of Liverpool and became part of its permanent collection. Inge Panneels and Jeffrey Sarmiento, artists both based at the National Glass Centre in Sunderland, won the competition proposing the *Liverpool Map*, a large sculpture composed of six columns of fused glass standing 2 metres high. Each column is made out of 17 layers of sheet glass, and each layer brings a different aspect of the geography, culture and history of the city, presenting a 'multi-layered snapshot of Liverpool in 2008' (Fig. 6.8).[26]

[25] Ibid.

[26] Davis, Laura; Shaw, Fiona; Panneels, Inge and Sarmiento, Jeffrey (2013). *Layers of Liverpool: mapping a sense of place. The story of the Liverpool Map*. Liverpool: Wordscapes, p. 82.

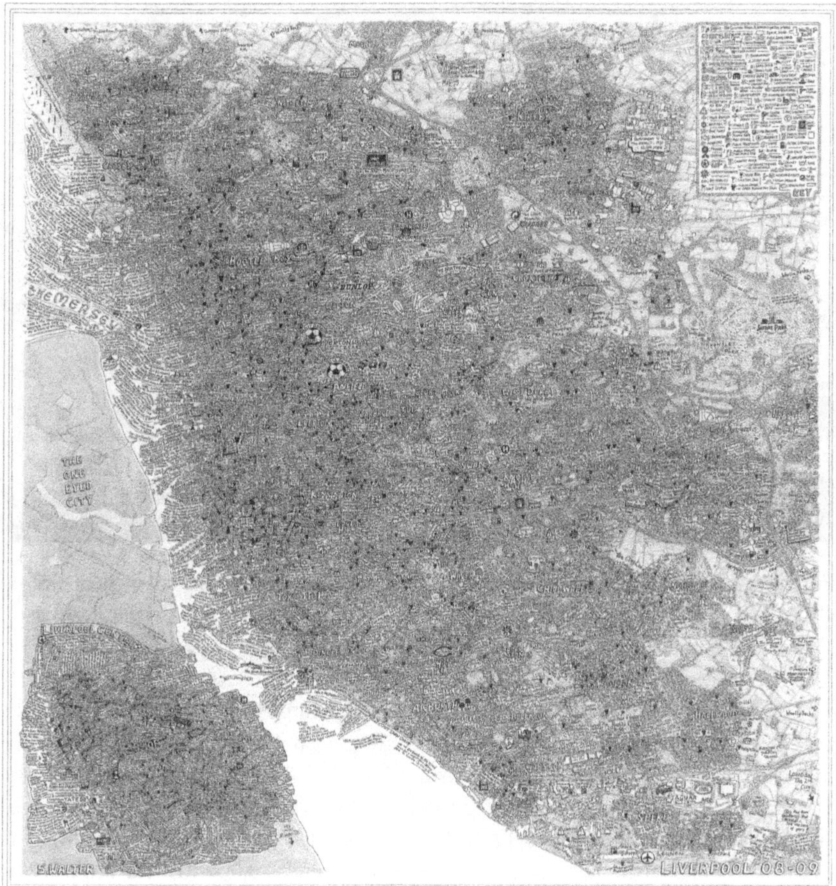

Fig. 6.7 *Liverpool 2008–09* by Stephen Walter. Inkjet print on paper 105 x 112 cm

Inge Panneels and Jeffrey Sarmiento are both practicing glass artists and academics, having been working at the University of Sunderland since 2006 supporting graduate and postgraduate courses in Glass and Ceramics at the National Glass Centre. Coming from different art practices—Panneels uses glass as a medium for site-specific artworks, while Sarmiento has a more technical approach, directed to produce gallery pieces—they pooled their efforts and their specific areas of knowledge to create the Liverpool Map in 2011.

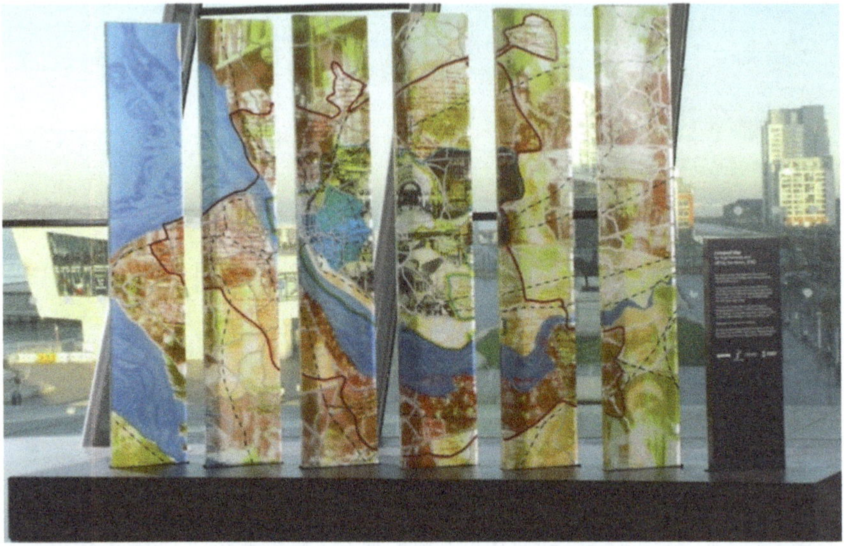

Fig. 6.8 The *Liverpool Map* by Inge Panneels and Jeffrey Sarmiento, as exhibited in the Museum of Liverpool

Inge Panneels was born in Belgium and moved to Scotland in the early 1990s, where she attended the Edinburgh College of Art. There she graduated in 1997 in Art and Design (Architectural Glass) and in 1998 completed a MA in Design (Sculptural Glass). Living and working in the Scottish Borders, where she set up her studio, she took up many public art and private commissions and produced work for exhibitions. In 2006 she became a part-time senior lecturer at the University of Sunderland while retaining her practice as artist. In 2014 she undertook a Doctorate at the Northumbria University investigating how artists are mapping climate change in the Anthropocene, a research project instigated by the making of the Liverpool Map. Currently she is a Research Fellow at Edinburgh Napier University, as part of the Creative Informatics project team.

Jeffrey Sarmiento was born in Chicago in 1974. While in the USA he graduated in Crafts and Art History and completed a MA in Fine Art Glass. Moving to England he followed a career as practicing artist and lecturer at the University of Sunderland, which he joined in 2006 and from where he received his PhD in 2011 studying expressions of ethnicity in glass artworks. A pioneer of print techniques for kiln glass, he has taught workshops in various countries in Europe and China, Australia, New Zealand and the USA. When working at the National Glass Centre, Jeffrey combined teaching and research, being Director of Studies for Postgraduate Research, as well as engaged in the Centre for Doctoral Training at Northumbria University. Currently he is Senior Lecturer at the School of Art and Design of the Australian National University.

The *Liverpool Map* reveals a unique profile of the city emerging from its many layers. Together they inform about the geography of the land, the urban origins, the transport links and the city's connection to the world (layers 1–4), the history of urban expansion and physical landmarks (layers 5–8), the social and emotional history of Liverpool (layers 9–12) and the personalities in music, sports, politics, culture and the arts that made the city known nationally and internationally (layers 13–16). One distinguishing feature in the map is the 'community layer' (layer 12), where extracts of the *Liverpool Saga*, an 800-line poem written by local people, were imprinted in glass. The poem was a project that stemmed from the celebrations of the 800th anniversary of Liverpool in 2007, when people were invited to write about the city's histories, cultural scenes and places that shaped their lives.[27]

The project was supported by Open Culture, an organisation created to involve people across the Merseyside region in the ECC cultural programme, and the National Museums of Liverpool. Behind those institutions was Phil Redmond, 'culture supremo of Capital of Culture year',[28] and main sponsor of the *Liverpool Map*, as well as Johnson's *Liverpool Cityscape*. Redmond and his wife Alexis offered financial support for both projects and were particularly closely involved in the production of the glass sculpture.

The piece was commissioned in 2008 but works did not start until 2009 and, as the artists described, it was "a true team effort" that received many contributions to its concept, development and production, which involved students at the National Glass Centre. As a glass artwork it was the first of its kind, because of the scale and the techniques employed, which combined precise waterjet cutting, screen printing and glass fusing.

[27] Poet Roger McGough wrote the eight opening and finishing lines, and the remaining 792 were selected among 3000 submissions sent from the public. Davis, Laura et al. (2013). Op. cit., p. 17.

[28] Davis, Laura et al. (2103). Op. cit., p. 23.

The work took almost a year to complete, and it was transported to the Museum of Liverpool for installation in 2011.

The three maps have overlapping narratives that refer to an imaginary that was constructed during the twentieth century in Liverpool, which has as its main components the music scene, football and architecture. The interpretation of this imaginary is different in each map though, which is partially due to the methods the artists employed to research and to collect information. None of the maps gives a comprehensive view of the city and its myths: as with any other map, they are built upon choices, selection and discernment. Borrowing from the concept used to produce the glass sculpture artwork, they are *layers* of the landscape and its social history, slices of the territory and urban life detached from reality and reinterpreted.

The artists engaged with the city and its inhabitants to produce their work through different degrees of interaction. All the artists are 'outsiders', in the sense that they are not from Liverpool or have never lived or worked there before, and for Inge Panneels and Jeffrey Sarmiento, it was the first time they visited the city. Stephen Walter was certainly the artist who had the most in-depth experience in Liverpool, being able to explore the city broadly during many excursions in various neighbourhoods in the city, talking to people from all walks of life and collecting first-hand accounts from them. Inge Panneels and Jeffrey Sarmiento conducted extensive research in the city's archives and received the input of people's choices and preferences to compose their map, absorbing a great amount of information on history and culture, but they did not have much of a personal experience and close contact with the city and the people. Ben Johnson's experience with the city was mediated by his own photographs during his visits to the city and other visual documentation; his only opportunity to reach out and establish a closer relationship with the public was during the time he was painting in the Walker Gallery.

Although dealing with the same city, the maps differ in terms of the actual territory covered: Johnson's panorama focuses on the city centre; it is a cropped view of Liverpool that is framed to include selected scenes. Walter's drawing goes as far as the urban landscape ends and the rural begins, making sure the entirety of the urbanised area is included. Panneels-Sarmiento's map has a city-region view, including the Wirral and further surroundings. If visually the maps maintain resemblances with traditional cartographical elements, as features of the landscape are placed with accuracy and the terrain is depicted to scale (Walter's and

Panneels-Sarmiento's maps are on scale, using OS data as a base), in terms of imagery and themes they are imprinted with strong local perceptions, transposing geographical boundaries and enacting subjective values, such as a sense of pride for the city and its people, a common attitude of resilience and defiance facing economic and social struggles, and the welcoming and friendly approach to strangers. The artists mentioned the expectations that maps induce in people, the importance that maps reflect peoples' places and histories and that, somehow, they want to be able to recognise themselves in the artwork, to identify and to use the map as a mirror for identity.

In the following sections I will discuss each artwork separately in the light of the interviews with the artists, examining in more detail the maps they created and comment on the artists' choices and their interpretations of the city within its historical and cultural context.

Portraying the Cityscape

Ben Johnson's cityscape is an architectural compendium of the city centre, a compacted view of its major iconic buildings. The Three Graces, the Museum of Liverpool and the Albert Docks dominate the foreground, followed by the Mann Island buildings and the Liverpool One retail complex just behind the museum, in evidence at the centre of the painting. In the background the most distinguishable features are the two cathedrals on the right-hand side, and the Radio Tower (known as St Johns' beacon) in conjunction with the Lime Street train station and St George's Hall, sided by the Walker Gallery, Central Library and the World Museum, composing the cultural 'acropolis' of Liverpool. The painting depicts the icons of the wealthy past of the city, that of the turn of the twentieth century, superimposed with the newly built icons of the twenty-first century on the waterfront. It is in many senses a nostalgic view that looks simultaneously to the past and to the future.

There is a common criticism around Johnson's work that points to the fact that his painting has no people on it. During my interview with Inge Panneels, she pointed out that 'most maps do not have people on them', joined by Jeffrey Sarmiento commenting that even cityscapes usually do not depict people, or if so, they are there for scale, to enhance a specific feature in the image. While both artists literally imprinted people on their *Liverpool Map*—peoples' writings and photographs of local celebrities—their comment is justified, as it is not expected to see people on maps. The public somehow participated in the process as observers but did not have a real influence on the final work. The map/panorama can be considered a view 'from above', akin to categorisations as 'traditional' or even 'authoritarian'.

What is interesting in this discussion is the problem of how to define and conceptualise maps: as a distant, detached view of the territory, a map does not allow the human presence; or better, the human presence is codified, as everything else, for the sake of representation. Having crossed at least five centuries of a quite well-established concept of the map, what is expected is that, to be recognised as such, a map more or less fits, or resembles, a cartographic standard of scale, projection and accuracy associated with a visual language.

Johnson's cityscape is firmly rooted in the Renaissance tradition of mapping. Taking the point of view from the river towards the urban skyline, the artist keeps up with the traditional urban profiles of the seventeenth and eighteenth centuries while revealing a contemporary eye in an unapologetic closeup of the city's most beautiful landmarks. However, dealing with the metropolitan reality of the contemporary city that long ago escaped the confines of its medieval walls, there is no attempt to depict the whole city—its boundaries are the boundaries of the canvas, and it is only in the background that one can realise where the city finishes and the countryside starts. Those margins are diluted, vanishing in the indistinct landscape. There is also no intention to restrict the image to an accurate layout of the city, but much more in evidence are the architectural landmarks that provide a sense of the 'spirit' of Liverpool; they are landmarks that everybody can recognise.

Johnson is clearly inspired by geometry and attracted to patterns, especially those used in Islamic art. He explained to me that he sees patterns not as decorative but 'a way of talking about something really quite profound and fundamental, like geometry - being a representation of an ideal world or even paradise on Earth'. This notion of the 'paradise on Earth' permeates Johnson's practice through his precise and meticulous method of painting, and even the colour palette reflects that, with a predominance of soft pastel colours, especially light grey, ochre and blue tones. Commenting on the hyperrealism of his paintings, Ben Johnson said:

> I think my paintings have a problem; they alienate a lot of people because they are rather too perfect. Certainly, I think a lot of people would criticise the painting saying 'Where is the dog shit, where is the dirt, where's the

> crime? Why have not you put any people on it? My answer to that - why have not you put any people on it - is because it is not really a painting of a real city, it is a painting of a model, of an ideal city. It's the city you might see in a dream.

This dream city is not that different from the one that can be seen in Google Maps: when comparing Johnson's 'imagined' interpretation of Liverpool with the 'real' 3D view captured from satellite imagery, the resemblance is striking, considering that this is a technology that was not available at the time of Johnson's painting. Google Earth was released in 2005 as a software combining satellite images with aerial photography, but it was only in 2009 that Google released the first bird's-eye city views of San Francisco Bay, in the USA, using 45° angle aerial imagery; ten years later the possibilities for navigation, zooming in and out and 'flying over' are extended to hundreds of cities around the world. Google's version of Liverpool might be more accurate in terms of scale and site positioning, but visually it is a rough picture: most of the buildings' outlines are not well defined, the colours are emaciated and the whole landscape looks covered by a bleak artificial light. It is just not visually appealing at all, and certainly it appears more devoid of life than Johnson's painting.

One can argue that the artistry reveals itself in making the illusion works. During our conversation we discussed the Renaissance's city views and the artists' pursuit of representing the real 'as it is seen', but even using accurate photogrammetry or satellite images, there are still choices to be made in mapping contemporary cities, as Ben Johnson comments:

> Yes, there are still decisions, and there are very simple decisions which can make an enormous difference. On computer models, for instance, you still 'have to observe it' and decide 'how to observe it'. Do you move around it or, if you want a single viewpoint, where do you stand? […] If you do a wide-angle view, you get a very distorted impression of how that city is. Telephoto is equally distorted, but possibly better for an embracing view.

In common with traditional techniques in producing bird's-eye views, Johnson commented that he also relies on 2D maps and buildings' footprints, from which he extrudes 3D views through perspective skills, in the past by hand and now using computers, although the technical concept has not changed much. As an example, Johnson describes his reconstruction of the fifteenth-century painting 'The Ideal City of Urbino':[29]

> I think it's quite clear that it is the work of somebody who understood perspective perfectly [...] when you take top view you can see that the perspective has been distorted enormously. So, in a computer model the squares at the back become rectangles and at the front they are a little bit less than squares, they are squashed in the opposite direction. That sort of manipulation fascinated me; that's why I made the painting' [...] Seeing it on plan it is again evidence that our maps can often bear distortions but sometimes you need to sort things to make them believable.

When asked why he uses these very laborious techniques on cityscapes that go back to traditions of representational art, Johnson admitted 'I've gone back five hundred years; I'm using rules of the Renaissance' (which can be justified by what was said earlier about the appeal of the Liverpool map):

> I think that there is a large number of people that are very suspicious of contemporary art, because they believe it is the emperor's new clothes. They look at it and they are not sure if they have been fooled. They stand in front of one of my paintings and I think they are genuinely moved by the amount of time that has been put into it.

With the approach of the year 2000, Johnson realised that his paintings so far were 'celebrating buildings or individual parts of buildings, and [...] certain architects', and he thought that he 'should start to look at cities and explore relationships between buildings. This came along with a time in the architecture world when 'people started to say 'individual buildings don't matter; it is more about the whole, the planning of our cities'. He then decided to try to make 20 paintings of 20 world cities:

> It would be a document [for the future] [...] I knew that I would make rather clean, cheerful, positive thinking views of any of those cities. I knew, when I was doing research into those cities, there was social injustice, poverty, dirt, crime, grime, but I did not want to mark that. I hoped that this movement would go beyond people like me; it would be poets, writers, philosophers, anthropologists, that also would start look to at cities. I really hoped the paintings would be just one small catalyst that could start other people discussing things.

[29] There are other two—Baltimore and Berlin. They are very similar, painted around the same period and with conflicted attribution; hence, the titles refer to the cities they depict.

Hong Kong was the first painting of the series, commissioned in 1995. It was based on drawings made from a vantage viewpoint in the city. He went there for the inauguration in 1997; with the idea in mind that his work could somehow trigger public debate about the city itself, he was quite disappointed. 'It was a total failure'. The painting was delivered and ceremoniously hung on a wall—'I think it is so high up people just think 'that's a nice big photograph'—and Johnson realised that it was used 'in the world of business and diplomacy to create a friendship between a commercial company and the incoming government [...] It was very good publicity for them'.

The next commission came from a foundation that approached Johnson and asked for a painting of a 'peaceful Jerusalem', where Jewish, Muslim and Christian quarters would be represented equally. Johnson built a computer model based on city plans and elevations of the buildings; he then applied a telephoto lens to it to show 'buildings in the background and the foreground without too much distortion'. This was the first time he used a computer model 'in a very thorough way'. The painting was finished in 2000, and the idea was to bring the Pope to Jerusalem to unveil it. However, the city museum refused to take responsibility for the safety of the painting as it saw the event turn into a political target. The painting was never exhibited in public 'apart from two months when it was shown in Liverpool in 2008'.

It was then that Johnson was contacted by Ann Bukantas, from the Walker Gallery. She went to London 'with someone who was just about to become head of the museum of Liverpool. They said: we would like you to make a painting that could become a symbol of Liverpool 2008'. Johnson did not want to deliver a painting that would just be put in a frame and go up in a museum and proposed that they would give him 'gallery space for two or three months to finish the painting in public'. He went to Liverpool to 'look around' and found a 'very different city' from his memories of the 1950s, when he was a child and was confronted with a 'very dark, very dirty, very depressed' landscape[30]:

[30] Later on the interview he mentioned that his first visit ever to an art gallery was in the Walker Gallery, on the year Peter Blake received the John Moore's prize [1961]. It made a huge impression on him as an art student; it changed the way he saw galleries and museums. 'I thought these were halls of grandeur, these museums. But they aren't, they are full of the work of people listening to the Beatles, wearing denim jackets and collecting badges'.

> So, I went up and what I had not remembered, and I was probably too young to appreciate it in the 1950s, were the people. They were just extraordinary. I thought well, even if I cannot find the ideal viewing spot, I want to make a painting for these people who are very proud of their city [...] the people of Liverpool drove me. There are some glorious buildings but there are also some very imperfect buildings, and some very imperfect social circumstances and conditions within Liverpool. But I presented a painting... I always say, 'Liverpool with no crime and no grime', which is not the real city.

In Liverpool Johnson went to 'every tall building' looking for a general view of the city, without success. 'I did not want to be God-like looking down; I wanted still to be part of the city'. He then became aware that a wooden block model, made around 1998–2000, was being restored:

> I spent two days flying around the model with a camera in a small tripod [...] then I photographed all the buildings that appear in the painting individually creating the basis of a computer model to which the details were superimposed.

During the time spent at the Walker Gallery finishing the painting, Johnson interacted with the public and collected interesting stories, as reported back to me:

> One day at the Walker, which was truly a humbling experience, somebody, an old lady came over to me and said... I was behind a barrier so I could get on with the work and she could come over and said... 'I just want to shake your hand because I love Liverpool, I was born here, and I always loved it. When I tell my friends from outside Liverpool, I live in the most beautiful city they come and say: 'what on the Earth it means? This is not beautiful, it is dirty, it is crowded, it is this and that'. I don't see it like that, I see it exactly the way you are painting. And I have heard they are going to make postcards so I can send my postcard and say 'this is my city'.
>
> There was also one man one day called me over and he said 'I live in Dublin, but my cousin lives in Liverpool. I was a brick layer, and I built the radio tower, I laid the stones for that. My cousin sent me a clipping... I wanted to come and see if you done my building properly because I was so proud, it was the best job I ever did.
>
> Once we were really late trying to get the train and the taxi driver said 'I have seen you in the Walker. I think you have done a good job on that painting'. We said thank you very much, but we are in a terrible hurry now'. He said, 'I will get you to the station' and he went down all a one-way street the wrong way. He said, 'I can do that'.

> One person was looking at it one day and called his wife and said, 'look at this' and in this painting his building is about this big [pinch fingers] and the windows are about this big. He said 'You see that window? That is where we conceived our first child.
>
> There was one man who came and said 'I just came from my hospital visit because I had a very high blood pressure. They are really thrilled with me because they said that I do not know what you were doing but your blood pressure came right down. I just said I come watch Ben Johnson painting at the Walker' He said: It is so peaceful when I am in there'.'

Ben Johnson's painting was budgeted at five hundred thousand pounds. Phil Redmond and his wife Alexis Redmond put down the first hundred saying 'if the artist doesn't start now and we wait until the money comes in it will never happen.'... 'I will write this much in, and we will see what we can do'. Johnson suggested to raise funding via subscriptions; they made different prints at different sizes and prices:

> [...] some went down a hundred pounds each and there was a large edition of those. Firms of solicitors, accountants, and the University put a lot of money in. It was really through subscription, and I do not think it cost the council anything at all, or the people of Liverpool.
>
> I try to avoid the question of money because it becomes tricky. Lots of people can justifiably turn around and say: half a million pounds? We could have built a nursery, or my mum could have been in a hospice or... [...] The money that came to this project maybe would not have gone to other projects. NML [National Museums Liverpool] have done [...] a critical study of the painting and what it is represented in terms of income to the city: number of visitors, how many people had to be employed to cope with these visitors, hotels, shops etc. It is actually paid for itself over and over again and the painting is used around the world to try and describe what the city is and tell some of the stories.

Johnson said that the Liverpool painting 'fitted in completely' with the commission's requirements of celebrating a cultural year:

> I use a language that is five hundred years old because it is a familiar language. You don't have to read a book about pictorial representation to understand perspective. The average person would say 'It is a proper picture, isn't it? I know what I'm looking at.

I comment that it is intriguing that a painting like the Liverpool cityscape—so realistic—has such a great appeal among the public and that probably a photograph would not reach the same impact. Ben Johnson replies:

> I think there are many reasons why a photograph would not work. However, complex the photograph is and however much collaging [...] however good the camera is, however much technology goes into it you will not know that somebody made it with love and respect.
>
> Part of my intention was to have respect for the people who built the city, but I ensured that through my own respect for what I can do [...] [Liverpudlians] can see right through people maybe because many of them had to survive on their wits. So, the reason whereby I could justify the painting is that I believe that piece of canvas was raw material, was painted on, and it is imbued with my own attention to detail and my passion for my craft and my attempt to celebrate a number of buildings, which are all shown in a very clean, clear, transparent light.

Ben Johnson justifies the fact that there are no people in his painting saying that 'I wanted anybody standing in front of it to be the only person that travels there'. The only concessions he made in the Liverpool map were the inclusions of the figure of Billy Fury and the war memorial, which were not planned at the beginning but were added at the request of members of the public visiting the Walker gallery (Fig. 6.9).

Narrating the City

Stephen Walter's drawing is an individual enterprise, where resourced historical documentation in old maps, photographs and history books is complemented by direct experience within the urban environment. Storytelling is a central theme in Walter's map of Liverpool, a quintessential portrait of the city composed from collected memories of the people he met and the places he visited. As with Johnson's panorama, Walter's map is charged with imagery that intends to present a specific concept of the urban, but contrary to Johnson's 'dream city', Walter's is 'dirty', mischievous and often funny, and Liverpool is depicted within the context of Britain's cultural and political scenario with a critical and opinionated eye.

Walter started making maps in 2002, but even before then he had a great interest in landscapes, and his first works were based on the familiar landscapes he often visited, especially the suburban areas of London, where he grew up, and Manchester, where he lived while attending university:

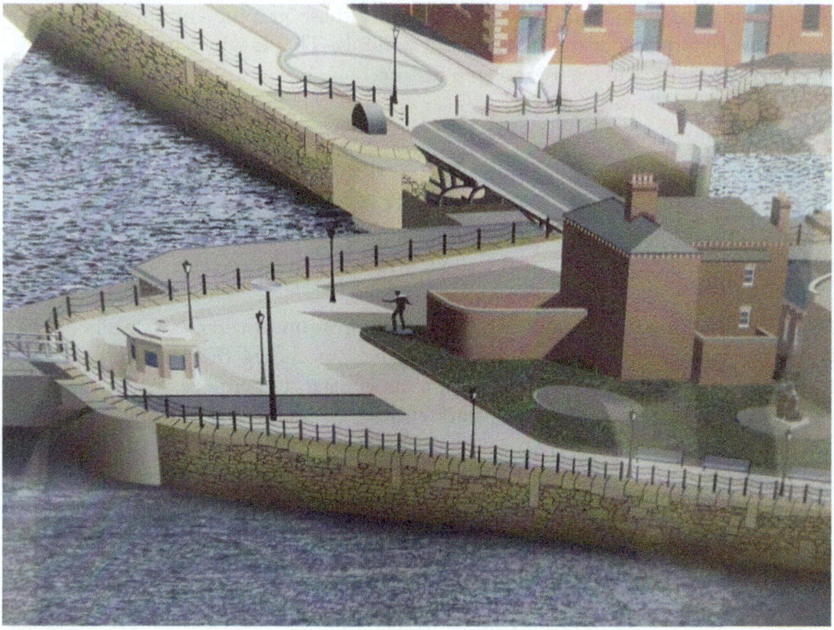

Fig. 6.9 *Liverpool Cityscape*—the Billy Fury statue (detail)

> The act of walking and observing is central in my practice. Often the springboard for any work is visiting a place myself, interacting with that meaning, personal meaning, as well as landscapes in general. [...] I have always loved traversing landscapes; it has always been part of what I have done. [...] The suburban environment of this post-industrial Britain was my playground, and that was where I started. It was always very important to concentrate on the areas I went to, even for the reason that I just lived there, so I recorded my journeys through the landscape memorialising certain viewpoints in drawings and paintings. My focus started in urbanised environments, then the fringes, and of course the very natural - the 'wilderness'.

This intimate relationship was built into Walter's work over the years through various techniques: walking, photographing, collaging, printmaking, drawing and painting. Registering what he observed and experienced turned later into a mark-making process using signs and symbols:

> I was interested in signs and symbols. Seeing them as things that have grown from a shared language of collective understandings. Not only are we shar-

> ing the landscape, but we are also sharing these graphic communications. Maps are landscapes full of signs and symbols. That is why I began to make maps. It seemed like a logical step for me.

These practices of registering the landscape are the background from where his work with maps emerged. Walter considers himself an 'artist of landscapes' and that his maps are expressions of what is imprinted in the landscape, of 'what we leave behind', a testimony of human endeavour, the stage where 'our physical rituals link with our desires':

> Maps are important because they are an expression of what we leave behind [...] I am into that idea of maps and landscape being the physical expressions, palimpsests [...] those documents constantly being re-written upon, and certain parts are from way back and are not rubbed out, but many are.

I asked if he considered maps as instruments for subversion and critique, in the way that situationists used them to challenge discourses about city and urban planning. Walter pointed that 'maps have been around ever since there have been humans' and he does not see that cartographic subversion is exclusive to situationists:

> I would not say mapping was their number one goal. They were talking about making a revolution of everyday life. Maps would be an obvious tool for that, but then maps have been an obvious tool for psychogeography since time immemorial. They always express the goals of their makers [...] maps have been spaces for subversion since they were made [...] If we look at early colonial maps in the 1600s: they were allied with a growing geographic knowledge, but ultimately, they were expressions of their growing influence and power upon the places that they had contact with. Maps would aid the colonialists in their growing structures of power and acquisition, but they always held the possibility of subversion.

In the contemporary world he considers that there is 'more space for subversions' and digital mapping is part of these practices:

> Postmodernism in a simple sense, is the decanting, dilating and breaking up of many things that once formed institutions. The focus has shifted towards ideas of the individual or the position of individual groups within the structure. Maps are part of this. [...] I think there is more ground for that dismantling of orthodoxy, it is part of what we have been talking about.

Another recurrent theme in Walter's maps is politics and the politics of space, and through his art practice, he expresses his concerns related to the use of natural resources and the contradictions of living in an over-populated world while dedicating his attention to unveiling the inequalities that political power exerts over the territory:

> I would not call myself an environmentalist, but it is one of the central pillars of concern in this kind of frugal practice that I do. I tend not to throw anything away, I regurgitate work, and my mapmaking is related to this [...] my maps have a sensibility towards environmental philosophy.

Making a parallel with the responsibility that architects and urban designers have in creating buildings and places on the landscape, he thinks that in the same way in his work he feels committed with ethics and politics of space:

> I am interested in that idea of responsibility, which of course as I said landscape is the stage on which all our lives get played. It is the physical manifestation of our ideas and goals and desires. [...] Maps are very much a comment on that. Even if you make a map of a private place, that private place is still part of the greater whole. Of course, now in a far too much overpopulated world politics of space is unavoidable. It is an intriguing and constant argument.

These ideas about the landscape as a source of inspiration and material to discuss human life also underpin the visual language adopted by Walter in his maps. Symbols and text were present in Walter's maps since his first experiments with them, but it was his map *The Island*, produced in 2008, that set his mapping practice in motion, leading to the development of the unique visual style and language that are recognisable in many of his artworks:

> [...] from 'The Island' it was just me diving into the mass, the wonderful interest in the city, and only just grasping a little bit of all the stories, that it says, revels in the chaos.

The Island was Walter's take on London, his home city, as the subject of a fictional place surrounded by its own made-up waters (as the 'Sea of Essex' and 'The Kentish Sea'), floating adrift somewhere in the world in an unrecognisable location on the globe. It is a playful depiction of

London as a cast-off place, which just reflects a common perception of the city as a country in itself that could thrive independently from the UK, as a city-state with its own culture and economy. The map keeps up with usual cartographic elements: it is to scale, it is placed with North up, it has an extensive legend with all symbols explained and it identifies features such as the river Thames, transport networks, built-up areas and the main parks with geographical accuracy. Overlaying the territory's backbone is a myriad of symbols and texts that relate to real and imagined venues, historical facts and myths, added to by Walter's own personal commentaries: his political stances against conservatism, his interests in folklore and pantheism, and his concerns with environmental problems that impact the landscape.

The map made Walter's work known; just after finishing it, he was approached by *The Guardian* newspaper, enquiring about the making of four maps of cities in the UK. At the time Walter agreed to go to Manchester and Liverpool for one week to undertake initial research. In his first stay in Liverpool he met a woman who was very knowledgeable about the history of the city; as they started a relationship Walter ended up getting into the 'veins' of the city. The newspaper commission never materialised, but Walter decided to go ahead and make a map for Liverpool by his own means:

> Initially it was about five or six days just going about paying cab drivers to show me suburbs here and there, just noting and picking up loads of information, like a private investigator of Liverpool. Then further research, and then about six months drawing. So I made a map of Liverpool, a sort of ode of love to Alexandra [...] Of course, I was also intrigued by how passionate the Scousers were about their city.

The symbols and the writing in Stephen Walter's map capture the dissonance and energy of Liverpool in numerous 'episodes' in the city's local history that are also much connected to national narratives. During the interview it was observed that maps are not impartial registrations of historical facts, people and places and they can reveal very personal political opinions, to which Walter commented:

> I certainly indulged my anti-Tory philosophy; it was very easy to let it play there! There was this playground song, a Margaret Thatcher song, something like clapping your hands and squashing her. Yes, things like that, and

of course the boycott of the Sun newspaper in the city [...] It was such a fascinating place to map.

If *The Island* was a vision of a hypothetical world made real through the fantastical and mythical, the *Liverpool 2008–09* map could be seen as a depiction of a real world made surreal and dreamlike. The absence of colour in the map emphasises a certain homogeneity given to the information, where there is no hierarchy or order, apart from a roughly correct geographical location. Walter's choice of symbols and texts combines funny and curious anecdotes alongside the crass, murderous or shameful facts. The information is placed according to places where they happened (or where they supposedly happened), but there is no attempt to give them much relevance over the whole context. Perhaps the most distinguishable symbols in the map when viewed from a distance are the two footballs that mark the Liverpool F.C. and Everton F.C. grounds. There is an inset of the city centre on the bottom left corner, but the visual aspect is the same as for the rest of the map.

The extensive legend, called 'key' in the map, follows a similar configuration: among pictograms for 'museum', 'hospital' and 'supermarket' are 'questionable fact (Wikipedia)', 'public hangings took place here' and 'reference to boobs or fit women'. These are all displayed in a continuous list, in contrast with being organised into groups, as is commonly seen in most city maps. There are other unexpected features in the legend, as for the housing categories: 'normal' (within inverted commas) domestic housing', 'rich people's home', 'town house', 'period house', 'stately home', 'high rise blocks & estate' and 'estate with satellite dishes'. This self-made classification evidently goes against the usual map categories, derived from field observation, and is probably very close to what a passer-by would spontaneously describe as a residential landscape, without much thought about the history or the demographics behind it (Fig. 6.10).

The mix of thoughtful research into the city's history and the artist's insights about what he sees around him creates a parallel urban universe, where an icon imprinted on the map acts as a gateway to other places and times. The map is an interpretation of the landscape in its physical aspects as much as a form of cultural storytelling. In the northern fringe of the city, for example, the coast of Crosby and Seaforth is 'littered' with the accounts of the local maritime history, such as the first ships that set sail from its shores or ships that brought in immigrants from Ireland. In the

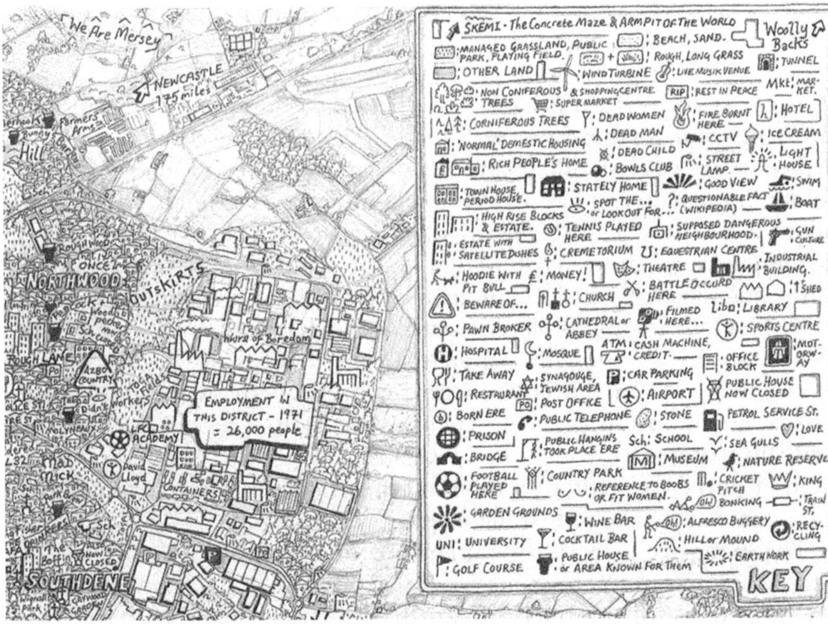

Fig. 6.10 *Liverpool 2008–09*. Detail showing the map's legend: going beyond the usual cartographic information

middle of the intricate writing of the dates and facts, one can read 'containerisation has meant that the Liverpool docks handles more cargo now than in its golden years: it is just the clanging and hum of machinery now, silencing the stories and wit of the dockers long gone'. Or in Lark Lane, a popular street with bars and cafes near Sefton Park, among the list of venues, it can be read 'where I got a gift from the people's poet'.

The juxtaposition of fact and fiction occurs in multiple directions: it intertwines events from different periods into the same location and also different mapping experiences, inscribing findings from desk research alongside observations of field work. Thus, the map plays with cartographic knowledge, challenging the notion of what is 'appropriate' to be on a map—not the neutral 'correct' history but the opinionated, site- and time-specific personal stories collected from the streets (Figs. 6.11 and 6.12). However, despite his map being overloaded with information, Walter considers that he only scratched the surface, in a critical self-reflection questioning about what it means to know a place:

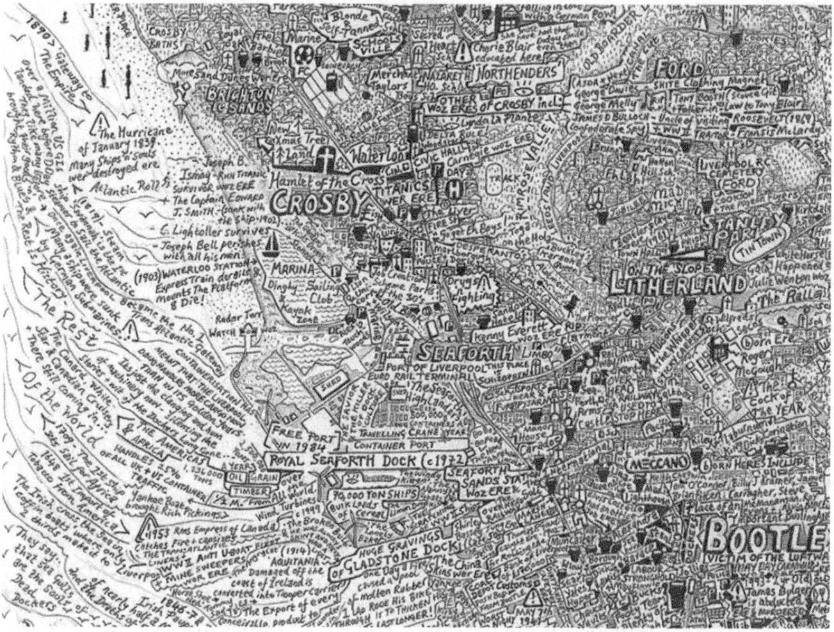

Fig. 6.11 *Liverpool 2008–09*. Detail of the north fringe of the city

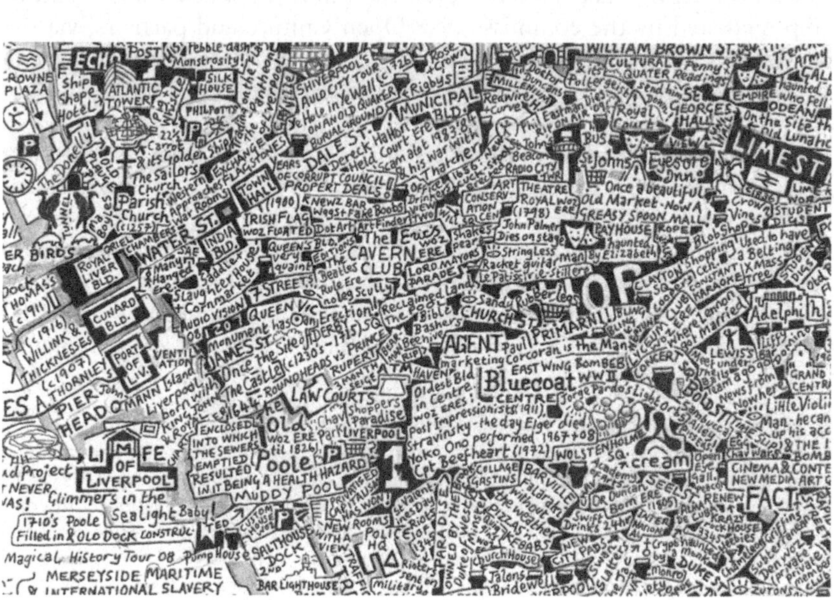

Fig. 6.12 *Liverpool 2008–09*. Detail of the central area of the city

Did I even really get to know Liverpool? You know quite superficial knowledge. There are many stories of Liverpool more relevant than what I did about Liverpool. I have drawn away over the last few years from mapping real places, unless it is London, which is my home and it always has been, I am not sure how profound I can be.

Mapping in Layers

From the first tentative scratch of the pen
to de keyboard's final breathless amen'
one poem. A patchwork of laughter and tears.
Eight hundred lines. Eight hundred years. Roger McGough, *Liverpool Saga*[31]

One main characteristic that distinguishes Panneels and Sarmiento's work from the other two maps is its participatory and collaborative production that permeates the mapping process from creation to execution. That involved the joint effort of two artists with different practices, the collaboration between the artists and their assistants and students at the National Glass Centre and the participation of the public in the composition of one of the map's layers—even though this last one was a relationship mediated by the commissioners Open Culture and partners, via the survey they collected.[32] This map is a result of a 'grounded' view of the city with the emphasis given to historical and geographical documentation, supporting the perspective of local inhabitants.

As a commissioned work the decision-making process was also the subject of intense participation of the sponsors. 'We had a quite vocal sponsor', Panneels said, and Sarmiento added that 'the level of engagement was quite high'. Therefore, the circumstances involving the production of the *Liverpool Map* in comparison with Johnson's painting were very different, starting with the fact that Johnson had already produced other similar works regarding cityscapes, while for the glass artists it was a completely

[31] Opening lines of the *Liverpool Saga* (2007), inscribed in the Liverpool Map.

[32] Open Culture was the organisation formed to bring together the cultural events and projects around the Capital of Culture year. Its mentor was Phil Redmond, chair of National Museums Liverpool and the main person behind the Liverpool Map. The National Museums Liverpool and Record Office provided the logistic support and information, while corporate partners Liverpool Daily Post and BBC Radio Merseyside took part in the *Liverpool Saga* and the public survey that decided which buildings, personalities and landmarks should be incorporated in the map.

new project, involving lots of experimentation and innovation. That meant that their work was transformed and taken in different directions through the process of design and execution:

> The first proposal was far more abstract, and it was only lately that it became so... geographically correct. The idea was always for the river to be one of the key features of the map. But it was only lately that it became quite so 'mappy', and we used the Ordnance Survey (OS) map as the baseline of this project. In that sense it is very much a map, a geographer's map. [Inge Panneels]

> We tried a few different ideas. Certainly, the early sketches were quite fun! Some of them were like grids, representing the blocks or sections of information, which is what you get if you split a map into various quadrants, or squares, or grids. He had the idea of making a folded map in glass, and we did actually go with that in one of the proposals. I think it was really cool, a crazy idea, but then it was technically impossible to make. [Jeffrey Sarmiento]

Inge Panneels and Jeffrey Sarmiento had the challenge to deal with a great amount of information from varied sources, including GIS databases, documents from the museum archives, historical maps, the input of people's writings and public survey. Specifically in relation to the GIS databases, Sarmiento appropriately noted that there was 'just too much information' and that 'at some point you have to eliminate it, because it would be unreadable' (Fig. 6.13). Visually, this process of 'cleansing'

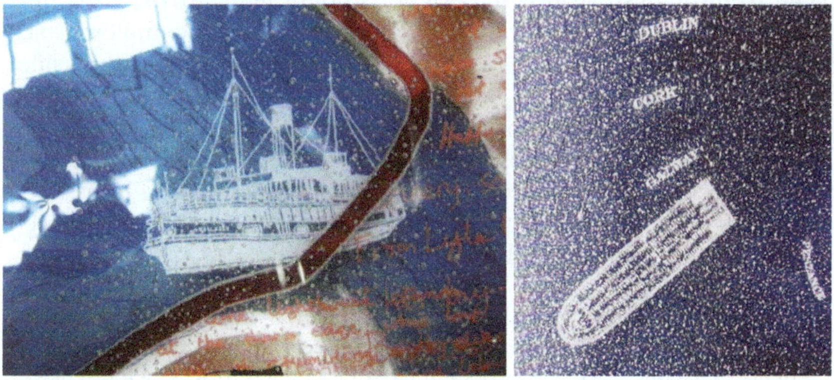

Fig. 6.13 *Map of Liverpool.* Detail of two of the many printed images generated from ink drawings. Photo: Inge Panneels/Jeffrey Sarmiento

information is equally important to communicate the message across, like in the case of the white outlined imagery printed in the glass, as Jeffrey Sarmiento reports:

> These are the ink drawings that were traced from images that already existed. In order to get the information you want to see from a photographic image, for instance, you have to either redraw it or modify it, and this is basically what we do: capturing the image, copying it, and then tracing on top of it, reducing information. Your source of information determines what your piece is going to look like in the end. [JS]

Despite this rich supply of information, the artists regret not having the opportunity to do more fieldwork and collect their own data, which was partly impeded by an unclear schedule:

> I think given the opportunity we would like to have spent more time in the city and do more 'under the skin' research. We were relying to a great extent on data that we were given, and we had certainly planned to do more primary research, but it took nine months to get the contract signed, and we were conscious of the time that we could dedicate to research when we were in limbo as to what was happening. [IP]

Both artists recognise that with hindsight they would do things differently, one of them being to use more colour. In fact, as the sculpture is on exhibition in the museum in front of a large window, some details are difficult to distinguish against the exterior's bright light. At the same time, they appreciate that they were able to develop such a complex project inside the National Glass Centre, benefiting from high-quality facilities and technical knowledge.

The *Liverpool Map* also had long-term impacts on the artists' careers, art practices and academic research. Jeffrey Sarmiento pointed out that the technical achievements proved to be important milestones in his research with glass printing and layering, which were incorporated into new techniques that came into use afterwards:

> It launched, for both of us, a lot of new investigations in the making of glass, the making of different artworks, which were deeply influenced by the experience. That one piece of artwork got us to a lot of new 'places': a lot of new technical territory, a lot of new artistic territory, I would say. And both of us are still using some elements of that. [JS]

For Inge Panneels the project was a catalyst for her further investigation into maps and artistic mapping practices, which led to her undertaking a PhD related to cartography and art. More interestingly the experience of this specific artwork resulted in the development of a course module at the University of Sunderland that targets mapping and mapping methodologies:

> The Liverpool Map forms a case study for our students at Sunderland. We bring some of that research, and some contextual developments of that, what a map is, what maps do, what is mapping in relation to site, place and space, and how they can be used as a tool. In that sense, our mapping practices as artists go right down to teaching. [IP]

The concept of the layers came from a technique of fusing glass sheets and playing with the transparency of the material, but it is also closely related to the way cartographers work, in digital platforms like GIS, superimposing categories of information. In the case of the *Liverpool Map* these categories could be identified into two main types: the physical geography, where the river is the central feature, and the local culture, dominated by the public's choices of personalities and landmarks. This last category has a temporal dimension that differs from the physical geography, as it depicts a very specific time in the history of the city, around the events of the European Capital of Culture:

> The map was both a collective and a personal response to the city of Liverpool, and obviously it was made for the occasion of the City of Culture. That was how people felt about the city at the time, it was only meant to reflect it. For example, the cultural icons of the time the map was made might be very different from what they are now. That will change, and in that sense, the map is very much a snapshot. [IP]

Assuming this particularity in the making of the artwork, and considering the influence of the sponsors in constructing a more 'mappy' layout for the artwork, it is fair to say that what is seen imprinted on the map is much a result of a time of celebration, and the people of Liverpool, without overlooking the struggles of the past, decided that the prized memories of success and achievements should be set in. There are marks of the distress, such as the ships that transported thousands of slaves in the era of transatlantic commerce, but the overall atmosphere is that of glorification.

On the other hand, the map does bring some interesting questions about what is to be understood as the city of Liverpool or what the identity of the city of Liverpool is, starting with the, supposedly, objective realities concerning the physical geography. Where does Liverpool start and end? Three boundaries are visible in the map: within the yellow line is the inner city, which also incorporates the river; the green boundary adjoins the Knowsley area and Huyton; and the red line embraces the Liverpudlian 'community', incorporating all the neighbourhoods and small towns that gravitate around the city of Liverpool. This includes the Wirral in the west (Wallasey, Birkenhead and Ellesmere Port), Runcorn in the southeast corner and Skelmersdale and Crosby in the north fringe:

> Asking local people 'What is the city of Liverpool?' was one the premises of the project. Someone from the Wirral, or someone from the north of Liverpool, would have a different opinion from someone living in the heart of the city. However they all feel connected to the city, and the biggest challenge for our project was to consolidate those opinions: that is why we ended up with the three lines. [IP]

The public was invited to vote on which places should fall within Liverpool's boundary lines, and some choices went as far as to include Southport and Chester,[33] but the final decision to trace the line had to be taken by the artists and commissioners, in order to give a coherent frame for the map that retained the most significant geographical and cultural aspects. This frame is clearly centred at the core of the city, expanding into all directions as if embracing the origins of Liverpool while contemplating the nearby settlements, connected through the river Mersey and the inland railways and roads (Fig. 6.14).

The images chosen for the map indicate a similar question—what does represent Liverpool? Again, as the public selected people, places and icons to be included in the map, the artists had to decide how to compose layers that would form a legible map, as Sarmiento suggested in his comments about how to withdraw useful information from the GIS databases. Concerning the city centre and main landmarks, there are many overlaps between what is imprinted in the *Liverpool Map* and Johnson's *Liverpool Cityscape*: the Three Graces, the Albert Dock, the Radio City tower and the two cathedrals appear in evidence in both maps. However, the

[33] For a complete list see Davis, Laura et al. (2013). Op. cit., p. 27.

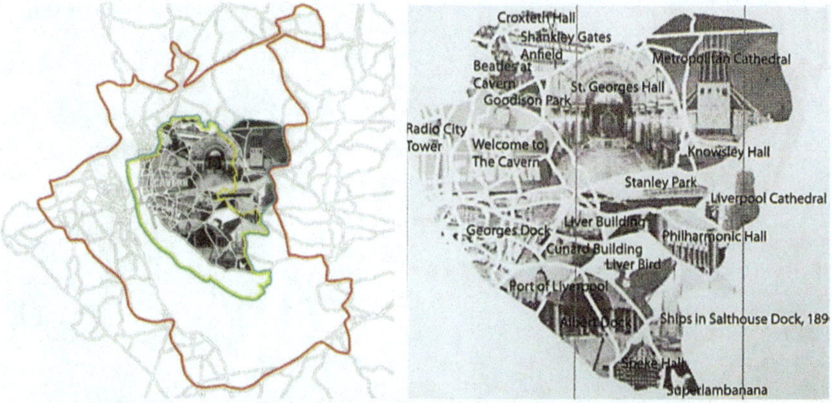

Fig. 6.14 *Map of Liverpool.* The boundaries underneath the map: the outer red line was suggested by Phil Redmond. On the right detail of the layer containing the iconic places of Liverpool. Photo: Inge Panneels/Jeffrey Sarmiento

extension of the *Liverpool Map*, going beyond the urban core, allowed the inclusion of other places that in Johnson's work are either not there or cannot be distinguished in the vanishing landscape in the background of the painting, such as Knowsley Hall, Croxteth Hall and Stanley Park. In addition, the arrangement of the layer as a collage of photographs made it possible to include places like the Cavern Club and single icons like the Superlambanana, a public art project installed in more than a hundred locations across the city during the celebrations of 2008's capital of culture.

The glass layering system offered flexibility to the artists to accommodate the request of the commissioners to integrate the information collected from the public, creating a visual language that amalgamates selected objects into common backgrounds that can be distinguished, or 'read', from the map (Fig. 6.15). In that respect the use of colour is a key element in the composition, as each colour indicates a different layer or thematic subject, producing an abstract picture that enables the map to function as a cultural portrait of the city, as Sarmiento explains:

> In the making of the Liverpool Map we were thinking about what would qualify as cultural. In my thesis I wrote that it was some kind of 'cultural terrain', some abstraction of terrain, using combinations of green and brown glass to develop what looks to be some form of relief, like what you find, in

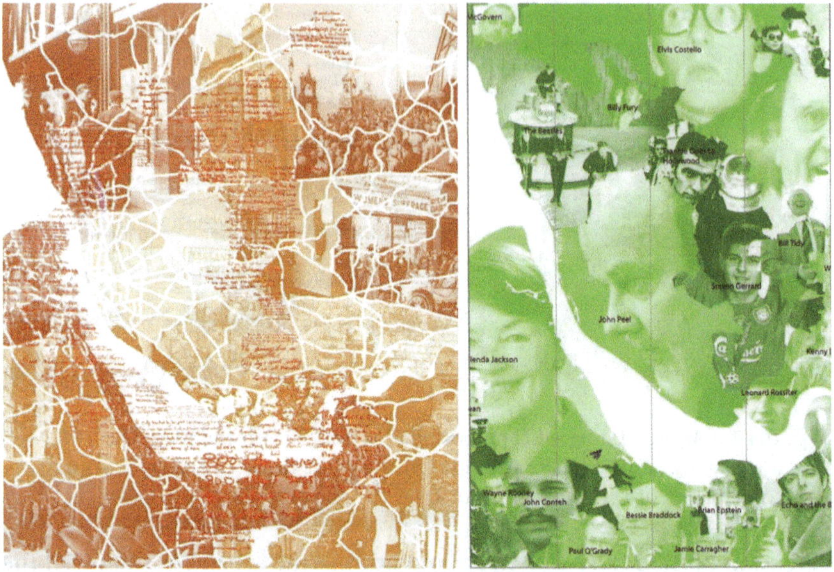

Fig. 6.15 *Map of Liverpool.* The community layer of people's writings superimposed on historical photographs (left) and the assembly of four layers depicting musicians, artists, writers, sports personalities and other people related to the cultural scene in Liverpool (right). Photo: Inge Panneels/Jeffrey Sarmiento

> terms of colours and colours choices, in a map. They kind of blend together, but they are actually made out of images as opposed to made of literal topographies or whatever that is. It is not a view form a high elevation as much as it is an indication of something that is local, and also from the ground. [JS]

Manipulating all the different sources of information and transforming data into an artwork can be seen as a work of translating concepts that define the map itself, as the artists pointed out:

> When you speak to geographers, or cartographers in particular, they have a very narrow view what a map is, and it is interesting to observe how people have different interpretations of maps. I have always argued that such definitions of maps can actually very easily be translated into a piece of art. [IP]

> What a lot of artists play on is trying to make something understood in a different way by someone else. So there is already a bit of translation happening, and that translation is something that artists capitalise on, the understanding or lack of it. All mapping is an interpretation; we manipulate maps to say what we want to say about place. [JS]

Epilogue: Viewing the Maps of Liverpool

On May 2019 I revisited the Liverpool maps in their exhibition venues, at the Museum of Liverpool and the Bluecoat. The glass sculpture was, reasonably, sitting in the same place where it was installed in 2011, at the right end side of the upper gallery on the second floor, as part of the People's Republic exhibition. The glass protected painting, however, had been moved from its original location where I first saw it, at the centre of the wall opposite the staircase on the first floor, to a more discreet back wall on the left-hand side of the same floor. The re-arrangement was necessary to free space for a temporary exhibition, I was informed. The framed hand drawing in the Bluecoat had also been moved, although it was a minimal change, transported from one end to another of the same curved wall in the entrance lobby.

The Museum of Liverpool lists Johnson's painting and Paneels-Sarmiento's glass sculpture as 'must-see objects' in its website, but the painting is clearly in the spotlight: it receives a longer description and more pictures, while only one paragraph is dedicated to the glass work, and which does not name the artists who created it. I observed the same imbalance in the Museum's gift shop, where Johnson's cityscape book, posters and postcards, along with decorated mugs, rulers and cushions, are available for purchasing, but nothing can be found related to Paneels-Sarmiento's artwork, not even their book. While viewing Johnson's painting on the first-floor gallery, I had a chat with a member of the staff in front of the big painting. 'Does it look like Liverpool?' I asked, to which she replied with a tilting of the head, 'somehow but' and then she went on pointing out 'mistakes' on it: things that do not exist anymore, or never existed; unfinished details, such as the tiles on the pavement behind the museum's building; and an interrupted fence. She found the scale odd, the ships were too small, the colour palette was 'very bland' and, most of all, there were no people. It looked artificial, like 'painting by numbers', and she preferred paintings where one can recognise the touch of a 'human hand' in the work. 'How about the glass sculpture on the second floor?

Do you like it?' I asked, curious to hear her opinion on a map full of people, but to my surprise she could not remember seeing that one, despite working in the museum since its opening.

Consulting the Bluecoat website I could not find a mention or pictures of Walter's drawing, as most of the contents focuses attention on the current programme of activities and exhibitions. When I entered the building, I walked straight to the spot where I used to see the map, to find a blank wall with some marks outlining the artwork's frame. Asking for help from a member of staff, it took her some time enquiring with her colleagues about the new location. During the search I heard that they could 'vaguely' remember having seen it 'somewhere'; the person in charge of the box office recalled: 'yes, we used to have prints here to sell but I have not seen them for a long time'. Upon finally discovering the drawing hanging on the same wall a few meters away, we were mesmerised by how it could be missed, hiding in plain sight.

I could not avoid a feeling of disappointment, as the maps seem to be forgotten, not attracting much interest neither from the public nor from the staff. Even in the case of Johnson's cityscape, with all the promotion that it still gets from the museum, it was clear that it did not have the same impact as when it was first created and exhibited. However, reflecting on this experience, I could make more sense of the domain of 'maps' and 'mappings'. The art maps are the repositories of the performative mappings that took place when they were produced, of that historical context and the experience of people who directly or indirectly participated in their making, from commissioners to anonymous contributors and audiences seeing, reading, observing and talking about them. The art maps as objects are the traces left from the creative enactment of the urban spaces of Liverpool conducted by the artists, and now they document such performances. In the same way that each art map came to materialisation through design investigation, geographical research, spatial analysis, fabrication and construction and many other aspects of their production, the objects we see in front of us, in those spaces of exhibition, reflect back on those processes.

CHAPTER 7

Placing/Performing the Map

A city incorporates uncountable tales, yet those that remain in the collective memory tend to be the ones registered and documented. Maps are powerful visual catalogues of the passage of time, and whenever cities came into evidence in history, at their origins, development and destruction, at times of great prosperity or acute impoverishment, they were mapped. Maps were also used throughout history to imagine cities-to-be, cities that were and have now disappeared, cities of fantasy worlds and cities of utopian and dystopian futures. Between registering and imagining, the creativity engendered in such mapping processes could not be dismissed by artists, and in the early twentieth century, a new cartographic tradition emerged in art practices, through means of performance, contestation and experimentation with maps and the territory. These early artworks are at the base of a contemporary map art that translates urban worlds as multiple realities, reinventing cities' spatial representations.

Cities have now been mapped and remapped countless times, yet they keep triggering the imagination of artists and cartographers alike as they are sites of intense and dynamic human relationships. In the twenty-first century, the ubiquity of digital maps and their widespread presence in everyday life, seem not to have affected the incubation of art maps and creative mapping processes, but instead to have nurtured them. Digital technologies expanded the cartographic realm, as much as the printing technology of the fifteenth century unleashed a lucrative map market.

G. Lanci, *Art Maps and Cities*,
https://doi.org/10.1007/978-3-031-13306-0_7

Similarly, these technological watersheds precipitated cultural changes, in terms of the way the world is narrated and interpreted. Contemporary artists mapping cities continue to play with the language and systems of cartography to express personal interpretations and alternative narratives of the urban world.

One of the aims of this book has been to question what contemporary art maps, as translations of urban spaces, communicate: what ideas of cities are behind them and how they are produced and exhibited. This was approached by interviewing artists and examining their artworks, and subsequent to this the problem of identifying 'artistic' from 'scientific' forms of map and mapping sprang up. As the research progressed it became clearer that I was dealing with a spectrum of practices, methods and epistemologies rather than two distinctive categories. On the one hand, it is not difficult to classify urban maps and 3D city models generated from the data manipulated in computer systems such as GIS as 'scientific'. On the other hand, it is more complex to separate the 'science' from the 'art' in other forms of mappings that combine elements of both sides, as two of the artists interviewed (Ruth Levene and Christian Nold) more explicitly discussed how these differentiations were challenged in their practices.[1] Therefore, it seems more relevant to point out that hybrid urban mappings are emerging and gaining significance in the context of the twenty-first-century contemporary art and that some artists are very aware of this tendency.

The other relevant aspect in new forms of artistic mapping is an increasing interest in bringing personal experiences *within/in* the urban environment (recalling Casey's categorisation,[2] presented in the Premise) to the foreground. Artists are incorporating mapping in their practices with the intention of revealing something beyond the exhaustive scrutiny of the purely 'informational' maps which, especially in the case of contemporary cities, are overwhelmed by a great amount of visual and statistical data. Some artists go through processes of mapmaking that are highly meticulous and laborious, employing painstaking techniques. This is better exemplified by the cases of the drawings of Kathy Prendergast and the photograph collages of Sohei Nishino, who both dedicated years of their careers working on their serialised projects, but this obsession with the

[1] As in Levene's As Far As The Eye Can See and Hidden Waters, and Nold's Emotion Maps.

[2] Casey, Edward (2005). *Earth-Mapping: Artists Reshaping Landscape*. Minneapolis: University of Minnesota Press, pp. xx–xxi.

urban matter, its configuration, morphology and landscape, can be also observed in the practices of Layla Curtis, Matthew Picton, Stephen Walter and Ben Johnson. Nevertheless, such artworks are displays of personal research and data gathering, and the 'information' mapped—being street networks, landscape snapshots, place names, history and stories, and architecture—are arranged and chosen by the artists according to their sensibilities and messages that they want to convey. They do not have to compromise with the depiction of an ultimate 'truth': instead, they can play with and question many 'truths'.

Certeau's dual metaphor of the 'voyeur'/'reader' and the 'pedestrian'/'writer' discussed in Chap. 3 must be brought up here again. Urban planners, sociologists and theoreticians tend to 'read' the maps and aerial views of cities, seeking structures and patterns of development that can lead to wider understandings of the complexities of the urban space and support analysis and intervention. Artists tend to engage with the process of 'writing' the city maps, inviting one to experience the urban space through selected themes and creatively explore multiple aspects of it. This does not mean that maps derived from the urban 'sciences' are depleted of the sensibility of first-hand human experience on the ground, but usually they do not generate an awareness of individual agency in the process of 'writing' the city, while the art maps stimulate this process, presenting an alternative that can link both ends of the spectrum, the analysed city and the experiential city.

Having said that, it must be considered that the authoritarian and top-down practices of urban planning and design, which, as was discussed in Chap. 2, remain in action in the age of the 'smart city' and the 'Internet of things', had since the 2000s been more often criticised and undermined by counter-actions of planning and mapping that incorporate art practices as well. Examples are found among the works of 'radical' geographers, cartographers and artists who joined forces to question official and institutional practices of surveying and mapping the territory.[3] As shown in Chap. 4, the political engagement of the arts with maps as conduits for action and transformation was already present in the works of Fluxus and conceptual artists of the 1960s and 1970s, through performance and

[3] See Mogel, Lize and Bhagat, Alexis (eds.) (2008). *An Atlas of Radical Cartography*, Los Angeles, Journal of Aesthetics & Protest Press; and Thompson, Nato et al (2008). *Experimental Geography: radical approaches to landscape, cartography and urbanism.* New York: Enfield: Melville House.

site-specific projects that, if not primarily oriented to protest and critique, were provocative exercises of cartographic disruption and disorientation.

The situationists in the 1950s developed methods of walking connected with mapping, but when minimalists, conceptual and land artists rebelled against the idea of the 'art as object' to emphasise 'art as experience', they opened the possibility for the art world to embrace mapping as a creative practice and move away from the fetishisation of the map as a disembodied object devoid of cultural values. Scholars in the early 2000s would identify the 'performative mapping' as the epistemological shift conducting more comprehensive and critical studies in cartography and acknowledging the intersections between artistic and scholarly activities.[4] Mapping as a performative enactment of space, as practice and method of embodied knowledge, revives Henri Lefebvre's theorisation of the urban space, grounded in the lived experience and corporeality of the individual, embedded in social practices.

During the course of researching this book it was particularly interesting to discover how artists respond to space and spatial practices in the most creative and varied ways, but in the case of mapping urban spaces, there is a tendency to play with the graphic and textual cartographic conventions of topographical maps. As examined in Chap. 3, topographical maps developed in the sixteenth century initiated a dominant tradition in city maps of orthogonal aerial views based on land surveys and symbolic representation of the terrain. This tradition is present today in commercial and institutional cartographic systems, both printed and digital. Despite the critiques of the gallery-artefact or the fetishisation of maps, the map-object is still a key component in discourses of cartography, of registering the land. It is an important aspect to build the register, not only because it guards the memory of the mapping but also it can act as a lever for alternative mappings and maps. It is a cycle of recreation of the narrative and reinvention of the space. At different levels of interaction, all the contemporary artists discussed in this project used ready-made city maps as informational and geographical reference or as base material for manipulation and transformation.

[4] Crampton, Jeremy (2009). Cartography; performative, participatory, political. Progress in Human Geography, 33 (6), pp. 840–848; p. 840; and Cosgrove, Denis and Martins, Luciana (2000). Millennial geographics. Annals of the American Association of Geographers. 90 (1) pp. 97–113.

The translational aspect of the artworks is more apparent in the challenges that artists face in creating maps that resemble cartographic pieces while at the same time personalising them as singular artistic interpretations of urban spaces. These challenges are related to constructing visual forms that can be 'readable' by the viewer as an artwork while simulating cartographic representations. General commentary or criticism that some artists received as not being 'scientific' (Christian Nold) or depicting 'true' reality (Layla Curtis and Stephen Walter) demonstrated the power of certain cartographic traditions and how the artists succeed in shaking them up. The inescapable materiality of their creations indicates that at the same time they are deliberately breaking cartographic rules they are also reorganising and recreating a new cartography, according to their own rules. Therefore, it is inevitable that translation is involved in communicating a 'foreign text'—in this case, one that is written with the traditional codes of cartography—by substituting it with renovated codes of artistic versions of the world observed.[5] In the case study of Liverpool, presented in Chap. 6, a double translation must be performed by the artists, who being strangers to that city have not only to provide their own artistic versions as well as to create a 'original text' reading from an urban space that they barely know. Stephen Walter, Inge Panneels and Jeffrey Sarmiento questioned the extension of knowledge that is produced through mapping a place that is unfamiliar to them. Walter implied that his work only scratched the surface of the history of the city and its people, while Panneels and Sarmiento recognised that their archival and desk research were not enough to develop a more in-depth knowledge of the city—also considering that they did not have the opportunity to do fieldwork.

While traditional cartographic conventions are often challenged, attempts to frame urban spaces within confined borders mostly disregarded, and this only emerged as an issue in the making of the Liverpool's glass map. Among the artists interviewed, Matthew Picton, for instance, opted for the circle as a form to limit the perimeter of some of his sculptures; others are delimited by roads, the map 'floating' against a blank canvas. The use of the street network to give a neat, clean edge to the map is also found in the works of Armelle Caron and Kathy Prendergast; more

[5] Following the reflections on translation by Venuti: 'Translation is the forcible replacement of the linguistic and cultural difference of the foreign text with a text that will be intelligible to the target language reader, Lawrence (1995). *The translator's invisibility: a history of translation*. London: Taylor & Francis, p. 18.

commonly the main subject, usually a central urban area, is framed within rectangular shapes, as a cut out of the territory, as in the case of Layla Curtis, Jeremy Wood, Sohei Nishino, Damien Hirst and Christian Nold. The maps of Liverpool are not different from this layout: in all of them the city centre is in the geometrical centre of the map, surrounded by neighbourhoods and the outer landscape. What is clearly distinguishable among them is the extension and scale of such frame: Johnson's cityscape is circumscribed to the core of the inner city; Walter's drawing extends to the borders of the urbanised area and the Panneels-Sarmiento sculpture generously incorporates the Merseyside region.

As cultural and social constructs cities are ever changing; and their physical reconfiguration is a continous and uninterrupted process. Any city map would be either obsolete as a representation of the past or unreal as a representation of the future, and the present cannot ever be captured in full. If on the one hand the mutable nature of the city defies cartographic science, on the other hand, it stimulates rhizomatic mappings, as suggested by Deleuze and Guattari,[6] infused simultaneously with official and imagined narratives. Urban change and transformation instigate multiple versions of urban reality, and as a city develops, its translations also change.

In the early twenty-first century, the physical reality of endless cities with no defined edges defies the very concept of 'urban'. There is no detectable unity prevailing in the urban space, and as capitalist urbanisation evolves, the trend is that urban form will continue to break apart. Henri Lefebvre argued that one of the reasons why an urban totality cannot be grasped is the fragmentation that the capitalist system imposes, fragments that are produced by gentrification, inequality, polarisation.[7] Taking Lefebvre's observation into a more general argument, it can be said that in a space of flux as dynamic as the city, the only constant is change. Thus, capitalist fragmentation is a strategy to keep portions unchanged and in doing so to sustain the system of cyclic consumption. Some artists interviewed, such as Ruth Levene and Stephen Walter, disclosed their concerns with the impact of rapid environmental changes in

[6] Deleuze, G. and Guattari, F. (1987) *A thousand plateaus: capitalism and schizophrenia*. Trans. by Brian Massumi. Minneapolis: University of Minnesota Press.

[7] 'The urban phenomenon can only be comprehended as a totality, but its totality cannot be grasped. It escapes us. It is always elsewhere'. Lefebvre, Henri (2003). *The urban revolution*. Trans. Robert Bononno. Minneapolis: University of Minnesota Press, p. 186.

human lifestyles and our survival on the planet, in tandem with other topics that are intersecting arts and the sciences and are leading artists to rethink maps.

Cities are the evidence of human efforts of processing the land, and they bear the legacies of human permanencies and transiencies over this planet. Exploring urban territories through mapping the urban form and life, contemporary artists are stimulating the debate of alternative narratives, making visible connections that otherwise would remain unexamined. Mapping practices that are questioning spatial truths and the act of mapping itself demonstrate that artistic intervention constructs a different form of knowledge. The type of artist/scholar collaborations that emerged in the twenty-first century, with the input of new digital technologies, also confirms that creativity is necessary to revitalise sciences and urban studies, beyond the comfortable stability of cartographic interpretations. For all the challenges that are ahead of city dwellers, which now count for more than half of the world population, threatened by social and environmental injustices, mapping differently matters more than ever.

Bibliography

Abrams, J., & Hall, P. (Eds.). (2006). *Else/where: Mapping new cartographies of networks and territories*. University of Minnesota Design Institute.

Alexander, C., Ishikawa, S., & Silverstein, M. (1977). *A pattern language: Towns, buildings, construction*. Oxford University Press.

Alpers, S. (1983). *The art of describing: Dutch art in the seventeenth century*. University of Chicago Press.

Altshuler, B. (2013). *Biennials and beyond: Exhibitions that made art history, 1962–2002*. Phaidon Press.

Arendt, H. (1958). *The human condition*. The University of Chicago Press.

Bachelard, G. (1969). *The poetics of space* (M. Jolas, Trans.). Beacon Press.

Badiou, A. (2004). Thèses sur l'art contemporain. *Performance Research, 9*(4), 86–86.

Ballon, H., & Friedman, D. (2007). Portraying the city in early modern Europe: Measurement, representation and planning. In D. Woodward (Ed.), *The history of cartography: Cartography in the European Renaissance* (Vol. 3). The University of Chicago Press.

Barthes, R. (1967/1997). Semiology and the urban. Lecture given on 16 May 1967. In N. Leach (Ed.), *Rethinking architecture: A reader in cultural theory* (pp. 158–164). Routledge.

Batey, P. (2011). The history of planning methodology. In C. Hein (Ed.), *The Routledge handbook of planning history* (pp. 46–59). Routledge.

Baudelaire, C., & Mayne, J. (Trans.). (1995). *The painter of modern life and other essays* (2nd ed.). Phaidon.

Bayley, S., & Liverpool City Council. (2010). *Liverpool: Shaping the city*. RIBA.

G. Lanci, *Art Maps and Cities*,
https://doi.org/10.1007/978-3-031-13306-0

Belchem, J. (2006). *Liverpool 800: Culture, character & history.* Liverpool University Press.

Belchem, J. (2007). *Irish, Catholic and Scouse: The history of the Liverpool-Irish, 1800–1939.* Liverpool University Press.

Bendall, S. (2002). Draft towns maps for John Speed's 'Theatre of the Empire of Great Britanie'. *Imago Mundi, 54*, 30–45.

Benjamin, W. (1999). *Illuminations: Essays and reflections* (H. Arendt, Ed. and H. Zohn, Trans.). Pimlico.

Benjamin, W. (1999). *The arcades project* (H. Eiland & K. McLaughlin, Trans.). Harvard University Press.

Berger, J. (1973). *Ways of seeing.* British Broadcasting Corporation and Penguin Books.

Bertin, J. (1983). *Semiology of graphics: Diagrams, networks, maps.* University of Wisconsin Press.

Biggs, I. (2010). Deep mapping: A brief introduction. In K. E. Till (Ed.), *Mapping spectral traces.* Virginia Tech College of Architecture and Urban Studies.

Blanchot, M. (1990). Translating. Trans. R. Sieburth. *Sulfur, 26*, 82–86.

Blazwick, I., & Institute of Contemporary Arts. (1989). *An endless adventure. An endless passion. An endless banquet: A situationist scrapbook.* Institute of Contemporary Arts.

Blazwick, I. (2001). *Century city: Art and culture in the modern metropolis.* Tate Gallery.

Bodenhamer, D., Corrigan, J., & Harris, T. (Eds.). (2010). *The spatial humanities: GIS and the future of humanities scholarship.* Indiana University Press.

Bodenhamer, D., Corrigan, J., & Harris, T. (Eds.). (2015). *Deep maps and spatial narratives.* Indiana University Press.

Brenner, N. (2009). What is critical urban theory? *City, 13*(2–3), 198–207.

Brenner, N., & Schmid, C. (2015). Towards a new epistemology of the urban? *City, 19*(2–3), 151–182.

Brinkman, S. (2014). Unstructured and semi-structured interviewing. In I. Leavy & Patricia. (Eds.), *The Oxford handbook of qualitative research.* Oxford University Press.

Brown, K. (2010). The artist as urban geographer. *American Art, 24*(3), 100–113.

Bruno, G. (2002). *Atlas of emotion: Journeys in art, architecture, and film.* Verso.

Bukantas, A. (Ed.). (2008). *Cityscape: Ben Johnson's Liverpool.* Liverpool University Press.

Busch, K. (2009). Artistic research and the poetics of knowledge. *Art & Research, 2*, 2.

Cairo, A. (2016). *The truthful art: Data, charts, and maps for communication.* New Riders.

Caquard, S., Vaughan, L., & Cartwright, W. (Eds.). (2011). *Mapping environmental issues in the city: Arts and cartography cross-perspectives.* New York.

Careri, F. (2002). *Walkscapes: Walking as an aesthetic practice*. Gustavo Gilli.
Casey, E. (2002). *Representing places: Landscape paintings and maps*. University of Minnesota Press.
Casey, E. (2005). *Earth-mapping: Artists reshaping landscape*. University of Minnesota Press.
Chtcheglov, I. (1953/2006). *Formulary for a new urbanism* (K. Knabb, Trans.). http://www.bopsecrets.org/SI/Chtcheglov.htm
Collingwood, R. G. (1938). *The principles of art*. Oxford University Press.
Comment, B. (1999). *The panorama*. Reaktion.
Corner, J. (1999). The agency of mapping: Speculation, critique and invention. In D. Cosgrove (Ed.), *Mappings*. Reaktion.
Cosgrove, D. (1994). Contested global visions: One-world, whole-Earth, and the Apollo space photographs. *Annals of the Association of American Geographers, 84*(2), 270–294.
Cosgrove, D. (1999). *Mappings*. Reaktion.
Cosgrove, D. (2001). *Apollo's eye: A cartographic genealogy of the Earth in the Western imagination*. John Hopkins University Press.
Cosgrove, D. (2005). Maps, mapping, modernity: Art and cartography in the twentieth century. *Imago Mundi, 57*(1), 35–54.
Cosgrove, D. C.-C., Abrams, J., & Hall, P. (Eds.). (2006). *Else/where: Mapping new cartographies of networks and territories* (pp. 148–165). University of Minnesota Design Institute.
Cosgrove, D., & Martins, L. (2000). Millennial geographics. *Annals of the American Association of Geographers., 90*(1), 97–113.
Couch, C. (2003). *City of change and challenge. Urban planning and regeneration in Liverpool*. Ashgate.
Crampton, J. (2009). Cartography; performative, participatory, political. *Progress in Human Geography, 33*(6), 840–848.
Crampton, J. (2010). *Mapping: A critical introduction to cartography and GIS*. Wiley-Blackwell.
Crampton, J., & Krygier, J. (2005). An Introduction to critical cartography. *ACME: An International Journal for Critical Geographies, 4*(1), 11–33.
Crary, J. (1998). Attention and modernity in the nineteenth century. In C. A. Jones & P. Galison (Eds.), *Picturing science, producing art*. Routledge.
Culcasi, K. (2015). Community mapping. In M. Monmonier (Ed.), *The history of cartography: Cartography in the twentieth century* (VI ed., pp. 286–287). The University of Chicago Press.
Curnow, W. (1999). Mapping and the expanded field of contemporary art. In D. Cosgrove (Ed.), *Mappings*. Reaktion.
D'ignazio, C. (2009). Art and cartography. In R. Kitchin, N. Thrift, & K. F. Al-Hindi (Eds.), *International encyclopedia of human geography* (pp. 190–206). Elsevier.

Daley, M. (1979). The debilitation of the British art school. *Art Monthly, 7*, 29–31.

Danto, A. C. (1964). The artworld. *Journal of Philosophy, 61*, 571–584.

Danto, A. C. (1997). *After the end of art: Contemporary art and the pale of history.* Princeton University Press.

Darroch, M., & Marchessault, J. (Eds.). (2014). *Cartographies of place: Navigating the urban.* McGill Queen's University Press.

Davis, L., Shaw, F., Panneels, I., & Sarmiento, J. (2013). *Layers of Liverpool: Mapping a sense of place. The story of the Liverpool map.* Wordscapes.

de Certeau, M.. (1984) *The practice of everyday life* (S. F. Rendall, Trans.). University of California Press.

De Quincey, T. (2008). The pleasures of opium. *Confessions of an English opium-eater* (Digital ed.). ProQuest LLC/Literature Online Prose.

Dear, M. (Ed.). (2011). *GeoHumanities: Art, history, text at the edge of place.* Routledge.

Debord, G. (1955\2006). *Introduction to a critique of urban geography* (K. Knabb, Trans.). http://www.cddc.vt.edu/sionline/presitu/geography.html

Debord, G. (1956). Theory of the Dérive. *Les Lèvres Nues* #9 (November) reprinted in *Internationale Situationniste* #2 (December 1958) (K. Knabb, Trans.). http://www.cddc.vt.edu/sionline/si/theory.html

Debord, G. (1972). On wild architecture. Preface to Ezio Gribaudo and Alberico Sala's *Jorn / Le jardin d'Albisola.* http://www.cddc.vt.edu/sionline/postsi/wild.html

Debord, G. (1994). *The society of spectacle* (D. Nicholson-Smith, Trans.). Zone Books.

Del Casino, V. J., & Hanna, S. P. (2006). Beyond the 'binaries': A methodological intervention for interrogating maps as representational practices. *ACME: An International E-Journal for Critical Geographies, 4*(1), 34–56.

Deleuze, G., & Guattari, F. (1987). *A thousand plateaus: Capitalism and schizophrenia* (B. Massumi, Trans.). University of Minnesota Press.

Dodge, M. (Ed.). (2011). *Classics in cartography: Reflections on influential articles from cartographica.* Wiley-Blackwell.

Dodge, M., Kitchin, R., & Perkins, C. R. (2011). *The map reader: Theories of mapping practice and cartographic representation.* Wiley-Blackwell.

Dorrian, M., & Pousin, F. (Eds.). (2013). *Seeing from above: The aerial view in visual culture.* I. B. Tauris.

Flam, J. (1996). *Robert Smithson: The collected writings.* University of California.

Foucault, M. (1997). The thought from outside. In B. Massumi (Trans.), *Foucault/Blanchot.* Zone Books.

Gardiner, M. (2000). *Critiques of everyday life.* Routledge.

Gould, P., & White, R. (1974). *Mental Maps.* Penguin.

Griffin, T., & Anastas, R. (2005). Method acting: The artist-interviewer conversation (with response). *Art Journal, 64*(3), 70–83.

Hall, P., & Tewdwr-Jones, M. (2011). *Urban and regional planning* (5th ed.). Routledge.
Harley, J. B. (2001). *The new nature of maps: Essays in the history of cartography* (P. Laxton, Ed.). John Hopkins University Press.
Harley, J. B. (1989). Deconstructing the map. *Cartographica, 26*(2), 1–20.
Harley, J. B. (1989). Maps, knowledge, and power. In D. Cosgrove & S. Daniels (Eds.), *The iconography of landscape: Essays on the symbolic representation, design and use of past environments* (Cambridge Studies in Historical Geography 9) (p. 292). Cambridge University Press.
Harley, J. B., & Woodward, D. (Eds.). (1987). *The history of cartography: Cartography in Prehistoric, Ancient, and Medieval Europe and the Mediterranean.* (Vol. 1). The University of Chicago Press.
Harmon, K. (2004). *You are here: Personal geographies and other maps of the imagination.* Princeton Architectural Press.
Harmon, K. (2009). *The map as art: Contemporary artists explore cartography.* Princeton Architectural Press.
Harvey, D. (1989). *The condition of postmodernity.* Basil Blackwell.
Harvey, D. (2001). *Spaces of capital: Towards a critical geography.* Edinburgh University Press.
Harvey, F. (2003). Knowledge and geography's technology—politics, ontologies, representations in the changing ways we know. In K. Anderson, M. Domosh, S. Pile, & N. Thrift (Eds.), *Handbook of cultural geography.* Sage.
Harvey, P. D. A. (1980). *The history of topographical maps: Symbols, pictures and surveys.* Thames and Hudson.
Hawkins, H. (2013). *For creative geographies: Geography, visual arts and the making of worlds.* Routledge.
Hein, C., & Mil, Y. (2020). Mapping as gap-finder: Geddes, Tyrwhitt and the comparative spatial analysis of port city regions. *Urban Planning, 5*(2), 152–166.
Hewitt, R. (2010). *Map of a nation: A biography of the Ordnance Survey.* Granta Books.
Hillier, B., & Hanson, J. (1984). *The social logic of space.* Cambridge University Press.
Hirst, D., Bracewell, M., Brotton, J., & Gallery, W. C. (2014). *Black scalpel cityscapes—Damien Hirst* (Exhibition Catalogue). Other Criteria.
Hlavajova, M. W., & J and Choi, B. (Eds.). (2008). *On knowledge production: A critical reader in contemporary art.* Utrecht.
Hollevoet, C. (1992). *Wandering in the city: Flânerie to dérive and after: The cognitive map of urban space* (Exhibition Catalogue). *The power of the city/ the city of power.*
Humpherys, A. (2002). Knowing the Victorian City: Writing and representation. *Victorian Literature and Culture, 30*(2), 601–612.
Ingold, T. (2000). *The perception of the environment: Essays in livelihood: Dwelling and skill.* Routledge.

Iuliano, M. (2009). Melchior Lorck's Istanbul in the European context. In E. Fischer (Ed.), *Melchior Lorck*. Vandkunsten Publishers.

Jacob, C. (2006). *The sovereign map: Theoretical approaches in cartography throughout history* (T. Conley, Trans.). (First published in French in 1992). University of Chicago Press.

Jacob, C. (1996). Toward a cultural history of cartography. *Imago Mundi, 48*, 191–198.

Jameson, F. (1991). *Postmodernism, or the cultural logic of late capitalism*. Duke University Press.

Jay, M. (1994). *Downcast eyes: The denigration of vision in twentieth century French thought*. University of California Press.

Jones, T. (2012). *Poetic language: Theory and practice from the Renaissance to the present*. Edinburgh University Press.

Kanekar, A. (2015). *Architecture's pretexts: Spaces of translation*. Routledge, Taylor & Francis Group.

King, G. (1996). *Mapping reality: An exploration of cultural cartographies*. Macmillan Press.

Kitchin, R., Gleeson, J., & Dodge, M. (2013). Unfolding mapping practices: A new epistemology for cartography. *Transactions, 38*(3), 480–496.

Knabb, K. (Ed. and Trans.). (2006). *Situationist international anthology*. Bureau of Public Secrets. http://www.bopsecrets.org/SI/

Korzybski, A. (1948). *Science and sanity: An introduction to Non-Aristotelian systems and general semantics* (3rd ed.). The Internal Non-Aristotelian Library.

Kulper, P. (2005). Representing beyond the surface. *arcCA 05.3—Drawn out*. American Institute of Architects, California Council (AIACC).

Lauriault, T., & Wood, J. (2009). GPS tracings—personal cartographies. *The Cartographic Journal, 46*(4), 360–365.

Lawton, R., & Lee, R. (Eds.). (2001). *Population and society in Western European port-cities, c. 1650–1939* (pp. 91–123; p. 115). Liverpool University Press.

Lefebvre, H. (1991). *The production of space* (D. Nicholson-Smith, Trans.). Blackwell.

Lefebvre, H. (2003). *The urban revolution* (R. Bononno, Trans.). University of Minnesota Press.

LeWitt, S. (2003). Paragraphs in conceptual art. In C. Harrison & P. Wood (Eds.), *Art in theory, 1900–2000: An anthology of changing ideas*. Blackwell.

Lodder, C. (2013). Transfiguring reality: Suprematism and the aerial view. In M. Dorrian & F. Pousin (Eds.), *Seeing from above: The aerial view in visual culture*. I. B. Tauris.

Lynch, K. (1960). *The image of the city*. MIT Press.

Lynch, K. (1984). Reconsidering the image of the city. In L. Rodwin & R. M. Hollister (Eds.), *Cities of the mind: Images and themes of the city in the social sciences*. Plenum Press.

Macarthur, J. (2013). On the obliqueness of the plan in urbanism and architecture. In M. Dorrian & F. Pousin (Eds.), *Seeing from above: A cultural history of the aerial view*. I.B. Tauris.

Macilwee, M. (2011). *The Liverpool underworld: crime in the city, 1750–1900*. Liverpool University Press.

Maher, S. N. (2014). *Deep map country: Literary cartography of the great plains*. University of Nebraska Press.

Maniglier, P. (2015). Maps in perspective; what can philosophy learn from experimental maps in contemporary art? In J. Levy (Ed.), *Cartographic turn*. Routledge.

Marcus, G. E., & Fischer, M. M. (1986). *Anthropology as cultural critique: An experimental moment in the human sciences*. University of Chicago Press.

Marin, L. (1988). *Portrait of the king* (M. M. Houle, Trans.). University of Minnesota Press.

Marin, L. (1984). *Utopics: Spatial play*. Macmillan.

Marin, L. (1993). Frontier of Utopia: Past and present. *Critical Inquiry, 19*(3), 397–420.

Marston, S., & De Leeuw, S. (2013). Creativity and geography: Towards a politicized intervention. *The Geographical Review, 103*, iii–xxvi.

Massey, D. (2005). *For space*. Sage.

McDonough, T. (2009). *The situationists and the city*. Verso.

McFarlane, C., & Soderstrom, O. (2017). On alternative smart cities. *City, 21*(3–4), 312–328.

McKee, F., & Irish Museum of Modern Art. (1999). *Kathy Prendergast: The end and the beginning* (Exhibition Catalogue). London.

Merleau-Ponty, M. (2002). *Phenomenology of perception*. Routledge Classics.

Millard, C. (2009). Kathy Prendergast in interview. *Circa, 128*, 42–51.

Mogel, L., & Bhagat, A. (Eds.). (2008). *An Atlas of radical cartography*. Journal of Aesthetics & Protest Press.

Monmonier, M. (1996). *How to lie with maps* (2nd ed.). The University of Chicago Press.

Monmonier, M. (2015). *The history of cartography: Cartography in the twentieth century* (Vol. VI). The University of Chicago Press.

Morgan, K. (2007). Liverpool's dominance in the British Slave Trade, 1740–1807. In D. Richardson, S. Schwarz, & A. Tibbles (Eds.), *Liverpool and the Transatlantic Slavery* (pp. 14–42). Liverpool University Press.

Mulhearn, C., & Franco, M. (2018). If you build it will they come? The boom in purpose-built student accommodation in central Liverpool: Destudentification, studentification and the future of the city. *Local Economy: The Journal of the Local Economy Policy Unit, 33*(5), 477–495.

Mumford, L. (1937/2011). What is a city. In R. T. Le Gates & F. Stout (Eds.), *The city reader* (pp. 91–95). Routledge.

Nesci, C. (2001). Flora Tristan's urban odyssey; notes on the missing flâneuse and her city. *Journal of Urban History, 27*(6), 709–722.

Newman, M., & Bird, J. (Eds.). (1999). *Rewriting conceptual art*. Reaktion.

Nold, C. (2018). Bio mapping: How can we use emotion to articulate cities? *Living Maps Review*, 4. http://livingmaps.review/journal/index.php/LMR/article/view/103

Noth, W. (1990). *Handbook of semiotics*. Indiana University Press.

Nuti, L. (1994). The perspective plan in the sixteenth century: The invention of a representational language. *The Art Bulletin, 76*(1), 105–128.

O'Rourke, K. (2013). *Walking and mapping: Artists as cartographers*. The MIT Press.

Osborne, P. (1999). Conceptual art and/as philosophy. In M. Newman & J. Bird (Eds.), *Rewriting conceptual art*. Reaktion.

Osborne, P. (2013). *Anywhere or not at all: Philosophy of contemporary art*. Verso.

Panofsky, E. (1991). *Perspective as symbolic form*. Zone.

Pascal, B. (2016). *Pensées*. Philosophical Library.

Peet, R. (2000). Celebrating thirty years of radical geography. *Environment and Planning A, 32*(6), 951–953.

Peluso, N. L. (1995). Whose woods are these? Counter-mapping forest territories in Kalimantan, Indonesia. *Antipode, 27*, 383–406.

Perkins, C. (2008). Cultures of map use. *The Cartographic Journal, 45*(2), 150–158.

Pickles, J. (2004). *A history of spaces: Cartographic reason, mapping, and the geo-coded world*. Routledge.

Pignatti, L. (Ed.). (2011). *Mind the map. Mappe, diagrammi e dispositivi cartografici*. Postmedia Books.

Pinder, D. (1996). Subverting cartography: The situationists and maps of the city. *Environment and Planning A, 28*, 405–427.

Pomian, K. (1998). Vision and cognition. In C. A. Jones & P. Galison (Eds.), *Picturing science, producing art*. Routledge.

Poole, R. (2008). *Earthrise: How man first saw the Earth*. Yale University Press.

Portugali, J. (2018). Cognition and the city: And introduction. *Built Environment, 44*, 2.

Reddleman, C. (2018). *Cartographic abstraction in contemporary art*. Routledge.

Ribeiro, D. M., & Caquard, S. (2018). Cartography and art. In J. P. Wilson (Ed.), *The geographic information science & technology body of knowledge*. https://doi.org/10.22224/gistbok/2018.1.4

Richardson, T. (Ed.). (2015). *Walking inside out: Contemporary British psychogeography*. Rowman & Littlefield.

Roberts, L. (Ed.). (2015). *Mapping cultures: Place, practice, performance*. Palgrave Macmillan.

Roberts, L. (Ed.). (2015). Deep mapping and spatial anthropology. Deep mapping. *Special Issue of Humanities, 4*(3), 569–599.

Roseau, N. (2013). The city seen from the aeroplane: Distorted reflections and urban futures. In M. Dorrian & F. Pousin (Eds.), *Seeing from above: The aerial view in visual culture*. I. B. Tauris.
Sadler, S. (1998). *The situationist city*. MIT Press.
Schuurman, N. (2000). Trouble in the heartland: GIS and its critics in the 1990s. *Progress in Human Geography, 24*(4), 569–590.
Sennett, R. (2008). *The craftsman*. Yale University Press.
Serres, M. (1982). *Hermes: Literature, science, philosophy* (J. V. Harari & D. F. Bell, Eds.). Johns Hopkins University Press.
Shepard, E. (1995). Sleeping with the enemy, or keeping the conversation going? *Environment and Planning A, 27*, 1026–1028.
Shiner, L. (2001). *The invention of art: A cultural history*. University of Chicago Press.
Simmel, G. (1971). The metropolis and mental life. In D. Levine (Ed.), *Simmel: On individuality and social forms*. University Press.
Soderstrom, O., Paasche, T., & Klauser, F. (2014). Smart cities as corporate storytelling. *City, 18*(3), 307–320.
Soja, E. (1989). *Postmodern geographies: The reassertion of space in critical social theory*. Verso.
Soja, E. W. (1996). *Thirdspace: Journeys to Los Angeles and other real-and-imagined places*. Blackwell.
Soja, E. W. (2000). *Postmetropolis: Critical studies of cities and regions*. Blackwell.
Spencer, J. R. (1966). *Leon Battista Alberti on painting* (Revised ed.). Yale University Press.
Swinson, J. (1999). Architects dream of nightmare cities. *Third Text, 13*(48), 101–106.
Sykes, O., et al. (2013). A city profile of Liverpool. *Cities, 35*, 299–318.
Thompson, N., et al. (2008). *Experimental geography: Radical approaches to landscape, cartography and urbanism*. Melville House.
Tibbles, A. (1996). Two paintings of Liverpool in 1680: A reassessment. *Transactions. The Historic Society of Lancashire and Cheshire, 146*, 189–195.
Tolman, E. C. (1948). Cognitive maps in rats and men. *Psychology Review, 55*, 189–208.
Tufte, E. (1983). *The visual display of quantitative information*. Graphic Press.
Velasco, A. d. P. (2008). Summary of 'An unknown object in uncountable dimensions: Visual arts as knowledge production in the retinal arena', a presentation by Sarat Maharaj. In M. Hlavajova, J. Winder, & B. Choi (Eds.), *On knowledge production: A critical reader in contemporary art*. BAK (basis voor actuele kunst).
Venuti, L. (1995). *The translator's invisibility: A history of translation*. Taylor & Francis.
Wark, M. (2011). *The beach beneath the street: The everyday life and glorious times of the situationist international*. Verso.

Watson, R. (2009). Mapping and contemporary art. *The Cartographic Journal, 46*(4), 293–307.
Wilford, J. N. (2002). *The mapmakers: The story of the great pioneers in cartography—from antiquity to the space age* (Revised ed.). Pimlico.
Wolff, J. (1985). The invisible flâneuse: Women and the literature of modernity. *Theory, Culture & Society, 2*, 37–46.
Wollen, P. (Ed.). (1989). *On the passage of a few people through a rather brief moment of time the Situationist International 1957–1972.* Institute of Contemporary Art.
Wollen, P. (1999). Mappings: Situationists and/or conceptualists. In M. Newman & J. Bird (Eds.), *Rewriting Conceptual Art.* Reaktion.
Wood, D. (1992). *The power of maps.* The Guilford Press.
Wood, D. (1993). The fine line between mapping and mapmaking. *Cartographica, 30*, 50–60.
Wood, D. (2006). Map art. *Cartographic perspectives, 53*, 5–14.
Wood, D. (2010). Lynch Debord: About two psychogeographies. *Cartographica, 45*(3), 185–200.
Wood, D.; with Fels, J. & Krygier, J. (2010). *Rethinking the power of maps.* The Guilford Press.
Wood, D. (2015). Mapping deeply. *Humanities, 4*(3), 304–318.
Wood, D., & Fels, J. (1986). Designs on signs/myth and meaning in maps. *Cartographica, 23*(3), 54–103.
Wood, D., & Fels, J. (2011). Reflection essay: Designs on signs/myth and meaning in maps. In M. Dodge & J. Crampton (Eds.), *Classics in cartography: Reflections on influential articles from Cartographica.* Hoboken, NJ.
Woodward, D. (Ed.). (2007). *The history of cartography: Cartography in the European Renaissance* (Vol. 3). The University of Chicago Press.

Index[1]

[1] Note: Page numbers followed by 'n' refer to notes.

G. Lanci, *Art Maps and Cities*,
https://doi.org/10.1007/978-3-031-13306-0

The manufacturer's authorised representative in the EU is Springer Nature Customer Service Centre GmbH, Europaplatz 3, 69115 Heidelberg, Germany. If you have any concerns regarding our products, please contact ProductSafety@springernature.com

Printed and bound by CPI Group (UK) Ltd, Croydon, CR0 4YY
28/09/2026
02224074-0003